Lecture Notes in Computer Science

Commenced Publication in 1973
Founding and Former Series Editors:
Gerhard Goos, Juris Hartmanis, and Jan van Leeuwen

Daniel Roggen   Clemens Lombriser
Gerhard Tröster   Gerd Kortuem
Paul Havinga (Eds.)

Smart Sensing and Context

Third European Conference, EuroSSC 2008
Zürich, Switzerland, October 29-31, 2008
Proceedings

 Springer

Volume Editors

Daniel Roggen
Clemens Lombriser
Gerhard Tröster
ETH Zürich, Wearable Computing Laboratory, Institut für Elektronik
Gloriastrasse 35, 8092 Zürich, Switzerland
E-mail: {roggen, lombriser, troester}@ife.ee.ethz.ch

Gerd Kortuem
Lancaster University, InfoLab 21, Computing Department
Lancaster, LA1 4WA, UK
E-mail: kortuem@comp.lancs.ac.uk

Paul Havinga
University of Twente, Department of Computer Science
P.O. Box 217, 7500 AE Enschede, The Netherlands
E-mail: havinga@cs.utwente.nl

Library of Congress Control Number: Applied for

CR Subject Classification (1998): H.3, H.4, C.2, H.5, F.2

LNCS Sublibrary: SL 5 – Computer Communication Networks and Telecommunications

ISSN	0302-9743
ISBN-10	3-540-88792-X Springer Berlin Heidelberg New York
ISBN-13	978-3-540-88792-8 Springer Berlin Heidelberg New York

Springer is a part of Springer Science+Business Media

springer.com

© Springer-Verlag Berlin Heidelberg 2008
Printed in Germany

Typesetting: Camera-ready by author, data conversion by Scientific Publishing Services, Chennai, India
Printed on acid-free paper   SPIN: 12547807   06/3180   5 4 3 2 1 0

Preface

This year marks the third edition of EuroSSC. It builds on the success of the past editions, held in Enschede, The Netherlands in 2006, and in Kendal, UK in 2007. On behalf of the Organizing Committee, we would like to welcome you to EuroSSC 2008, in Zurich, Switerland. This volume contains the invited papers and technical peer-reviewed papers selected for presentation at the conference.

At EuroSSC we aim to explore technologies, algorithms, architectures, protocols, and user aspects underlying context-aware smart surroundings, cooperating intelligent objects, and their applications. Since its inception, EuroSSC has taken a complementary technology-driven and user-driven view to discuss these aspects. It is one of the particularities of EuroSSC, and the 2008 edition made no exception. In addition we emphasized aspects related to quality of context and context-aware feedback by actuator systems. This reflects the growing importance that context processing in uncertain environments and sensor and actuator networks take in ambient intelligence environments.

We received 70 paper submissions. They originate from 30 countries of Europe, the Middle East and Africa (66%), Asia (22%), North America (9%), and South America (3%). These numbers reflect the European origins of EuroSSC, but also show that EuroSSC is a recognized and attractive platform for participants from all regions of the world.

The review process was double-blind. This ensured that the papers were selected purely on their technical merits. Each paper received a minimum of two reviews, with most receiving three reviews and some four. On the basis of the peer reviews, the Program Co-chairs selected 17 papers for publication. The 25% acceptance rate indicates the high quality of the papers included in this proceedings volume. The papers were selected according to their technical merit to exemplarily discuss the key themes addressed by EuroSSC: sensors and signal acquisition; context inference; high-level context processing; and applications and scenarios.

We invited Alain Crevoisier, researcher at the University of Applied Sciences, Yverdon, Switzerland, to give an invited paper presentation. Sensing being the first requirement of context-aware systems, he discussed how daily life objects can be turned into tactile interfaces, on the basis of his activities in the development of new musical instruments, notably within the EU-funded research project 'Tangible Acoustic Interfaces for Computer–Human Interaction'.

The third edition of a conference can arguably be considered a turning point. This year saw a 35% increase in paper submissions since the last edition. We are satisfied to have received contributions from both Americas, and a steady but very significant contribution from Asia. This marked increase in submissions and the increased diversity in the countries of origin (30 countries in 2008, 21 in 2007) shows that the conference has grown more attractive. We are confident that 2008 will be a successful milestone that will lead to future editions equally (and more!) successful.

A peer-reviewed conference cannot exist without a community supporting it. As a conference is a place to share and discuss ideas, we would first like to thank all the 179 authors who took time to write and submit papers. We are extremely grateful to the Program Committee and the additional reviewers who dedicated a lot of their valuable time to read, comment, rate and suggest ways to improve those papers. The peer-reviewing process is time consuming, but it is essential to ensure high-quality publications, and to provide constructive feedback and advice for the papers that could not be accepted. We would like to thank the authors of accepted papers for revising their papers according to the reviewers' comments. We hope the review process helped those authors whose paper could not be accepted. In order to provide an opportunity for these authors to present their work, we organized a poster paper session, with a delayed submission deadline, published in an adjunct proceedings with its own ISBN number.

We are thankful to the supporting organizations that helped us with EuroSSC: the Institut für Elektronik and ETH Zürich that supported this conference by providing the infrastructure, the Springer LNCS staff, the IEEE Communications Society for their technical co-sponsorship, and the EU project SENSEI and Sensinode Ltd. for their financial support. Our gratitude also extends to the numerous volunteers who helped us during the organization of this conference.

August 2008   Daniel Roggen
                                                    Clemens Lombriser
                                                     Gerhard Tröster

Organization

EuroSSC 2008 was organized and supported by the Wearable Computing Laboratory, Institut für Elektronik, ETH Zürich, Switzerland.

General Chair

Daniel Roggen   ETH Zürich, Switzerland

Program Chairs

Daniel Roggen   ETH Zürich, Switzerland
Clemens Lombriser   ETH Zürich, Switzerland
Gerhard Tröster   ETH Zürich, Switzerland
Gerd Kortuem   Lancaster University, UK
Paul Havinga   University of Twente, The Netherlands

Local Chair and Publicity Chair

Clemens Lombriser   ETH Zürich, Switzerland
Andreas Bulling   ETH Zürich, Switzerland
Ruth Zähringer   ETH Zürich, Switzerland

Poster and Demo Chair

Andreas Bulling   ETH Zürich, Switzerland

Finance Chair

Ruth Zähringer   ETH Zürich, Switzerland

Program Committee

Martin Bauer   NEC Europe, Germany
Xiang Chen   Institute for Infocomm Research, Singapore
Pham CongDuc   University of Pau, France
Jessie Dedecker   Vrije Universiteit Brussel, Belgium
Simon Dobson   University College Dublin, Ireland
Falko Dressler   University of Erlangen, Germany
Martin Elixmann   Philips Research, Germany
Elisabetta Farella   University of Bologna, Italy

Ling Feng	Tsinghua University China, P.R. China
Alois Ferscha	Johannes Kepler University of Linz, Austria
Elgar Fleisch	ETH Zürich, Switzerland
Kaori Fujinami	Tokyo University of Agriculture and Technology, Japan
Sandeep Gupta	Arizona State University, USA
Manfred Hauswirth	DERI Galway, Ireland
Paul Havinga	University of Twente, The Netherlands
Julia Kantorovitch	VTT, Finland
Gerd Kortuem	Lancaster University, UK
Marc Langheinrich	ETH Zürich, Switzerland
Rodger Lea	University of British Columbia, Canada
Peter Leijdekkers	University of Technology Sydney, Australia
Xinrong Li	University of North Texas, USA
Maria Lijding	University of Twente, The Netherlands
Paul Lukowicz	University of Passau, Germany
Oscar Mayora	CREATE-NET, Italy
Stefan Meissner	University of Surrey, UK
Nirvana Meratnia	University of Twente, The Netherlands
Tatsuo Nakajima	Waseda University, Japan
Santosh Pandey	Cisco Systems, USA
Christian Prehofer	Nokia Research, Finland
Kay Römer	ETH Zürich, Switzerland
Kamran Sayrafian-Pour	NIST, USA
James Scott	Microsoft Research Cambridge, UK
Frank Siegemund	European Microsoft Innovation Center, Germany
Junichi Suzuki	University of Massachusetts, USA
Hong-Linh Truong	Vienna University of Technology, Austria
Kristof Van Laerhoven	Darmstadt University of Technology, Germany
Jamie Ward	Lancaster University, UK

Additional Reviewers

Bashar Altakrouri	Lancaster University, UK
Hilbrandt Baarsma	University of Twente, The Netherlands
Konssalya Balasubramanium	Cisco, USA
Davide Brunelli	University of Bologna, Italy
Supriyo Chatterjea	University of Twente, The Netherlands
Isabel Dietrich	University of Erlangen, Germany
Daniel Fitton	Lancaster University, UK
Siang Fook Foo	Institute for Infocomm Research, Singapore
Xiangpeng Jing	Rutgers University, USA

Hui Kang	State University of New York at Stony Brook, USA
Santosh Kulkarni	Auburn University, USA
Michele Magno	University of Bologna, Italy
Raluca Marin-Perianu	University of Twente, The Netherlands
Feki Mohamed Ali	Institute for Infocomm Research, Singapore
Ilkka Niskanen	VTT, Finland
Abhishek Patil	Sony Electronics, USA
Romelia Plesa	University of Ottawa, Canada
Aameek Singh	IBM Almaden Research Center, USA
Agustinus Borgy Waluyo	Institute for Infocomm Research, Singapore
Wendong Xiao	Institute for Infocomm Research, Singapore
Piero Zappi	University of Bologna, Italy

Sponsoring Institutions

| Silver sponsors | European Research Project SENSEI |
| | Sensinode Ltd. |

IEEE Communications Society co-technical sponsorship

Table of Contents

Context-Aware Interaction and Case Studies

Transforming Daily Life Objects into Tactile Interfaces

Alain Crevoisier and Cédric Bornand

University of Applied Sciences Western Switzerland (HES-SO),
School of Engineering and Business Management Vaud (HEIG-VD)
Rue Galilée 15, CH-1400 Yverdon-les-Bains
{alain.crevoisier,cedric.bornand}@heig-vd.ch

Abstract. This article describes a few techniques to transform daily life objects into tactile interfaces, and presents the implementation details for three objects chosen as example: a light globe, a tray and a table. Those techniques can be divided in two main categories, acoustic techniques and computer vision techniques. Acoustic techniques use the vibrations that are produced when touching an object and that are propagating through and on the surface of the object until reaching piezo sensors attached on the surface. The computer vision approach is an extension of the technique used for virtual keyboards, and is based on the detection of fingers intercepting a plane of infrared light projected above the surface by a pair of laser modules. It allows for multi-touch sensing on any flat surfaces.

Keywords: Tangible Acoustic Interfaces, Acoustic Sensors, Computer Vision, Signal Processing, Computer-Human Interfaces.

1 Introduction

In relation with the development of ambient intelligence and pervasive computing, there is a need to create intelligent objects, as well to interact with our surrounding environment in the most natural and intuitive manner. In this context, our interest focused since several years in finding ways to use existing objects as human-computer interfaces. Among the various information that can be retrieved from an object and used in interactive contexts, such as the position, orientation, size, etc, our attention concerned mainly the sense of touch. Our goal was that such objects could be transformed into tactile interfaces with a minimum of modification. This reduces dramatically the scope of suitable sensing methods. Indeed, most touch sensing methods require a sensitive layer to be applied on top of the surface one wants to make tactile (eg. touch screen or touch pad). Moreover such layers require the surface to be flat, while a large number of our surrounding objects are curved. This led us to consider primarily acoustic sensing technologies [6] and, more recently, computer vision technologies. Acoustic sensing has the advantage of integrating smoothly with an object and with a minimum of intervention (only a few sensors to glue on the surface), but is more subject to perturbation due to the manipulation of the object or due to ambient noise. Also, continuous tracking of touch (eg. dragging a finger) is particularly complex and still limited to thin and flat objects. On the other hand,

D. Roggen et al. (Eds.): EuroSSC 2008, LNCS 5279, pp. 1–13, 2008.

computer vision techniques allow for tracking easily continuous movements but make difficult to detect when touching or not an object. Chapter 2 and 3 will give an overview of the various acoustic and computer vision techniques we have been involved with, either as lead researchers or in collaboration with other research teams in the context of European or national projects. Chapter 4 will provide more details of the practical implementation for three object examples: a light globe, a plastic tray and a table.

2 Acoustic Techniques

Tactile interfaces based on acoustic detection techniques usually refer to the name of Tangible Acoustic Interfaces (TAIs). Tactile information is determined by the mean of the acoustic vibrations that are produced when touching or manipulating an object, either with the hand or with another object (e.g. a stick or a pen). By analyzing those vibrations, it is possible to determine *where* and *how* the object is touched, thus providing a complete description about the tangible interaction.

Many other touch technologies exist but their common disadvantage is either the presence of mechanical or electronic devices at the point(s) of interaction (switches, potentiometers, sensitive layers, force resistive sensors, etc), or the necessity to use a specific kind of material (acrylic pane or semitransparent film) in case of screen based touch interfaces [4], [5]. The major advantage of TAI's is their ability to use any kind of material that can transmit acoustic vibrations, such as metal, wood, plastic and glass, and without the need of fitting the object with intrusive sensors or devices in the area of contact and interaction. This opens the door to transform daily life objects and surfaces into interactive interfaces.

First experiments about TAI's took place at MIT in the 90's [1]. Further researches have demonstrated potential applications for building large-scale interactive displays [2], [3], and for creating new musical interfaces [6], [7]. More recently, a European project of research (TAI-CHI project) has embraced the subject more widely, leading to new technology breakthrough, such as the continuous tracking of fingers touching a surface or the use of 3D objects [8].

There is no single TAI technology, but an ensemble of various techniques and approaches. In the following sections, we will concentrate on two techniques we have been directly involved with and used in two of the examples presented in chapter 4.

2.1 Time Reversal

Time reversal in acoustics is a very efficient solution to focus sound back to its source in a wide range of material including reverberating media [9]. It is based on the principle that the impulse response in a chaotic cavity is unique for a given source location. The method that is employed here is a particular case of the Time Reversal technique and consists in detecting the acoustic waves in solid objects generated by a simple human touch. The detection is a two steps process. The first one is the acquisition of the impulse response: a short pulse is emitted by tapping on the surface of the object, which propagates toward the solid cavity and reflects inside. The reflections are collected by a contact transducer working as a receiver (Figure 1 - Top). The duration of the response depends on the absorption of the material and on the energy radiation property of the cavity.

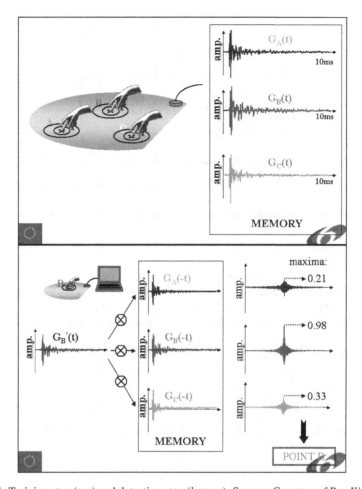

Fig. 1. Training step (*top*) and detection step (*bottom*). Source: *Courtesy of Ros Kiri Ing.*

In the second step (Figure 1 - Bottom), the information related to the source location is extracted by performing a cross-correlation between the stored signals and the live input. The number of possible touch locations at the surface of an object is directly related to the mean wavelength of the detected acoustic wave [10].

2.2   Time Delay of Arrival (TDOA)

TDOA-based locators are all based on a two-step procedure applied on a set of spatially separated microphones. Time delay estimation of the source signals is first performed on pairs of distant sensors (Figure 2). This information is then used for constructing hyperbolic curves that describe for each couple of sensors (the foci of the hyperbola) the location of all points that correspond to the estimated delay. The

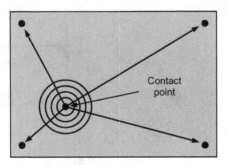

Fig. 2. Basic principle of TDOA estimation

curves drawn for the different pairs of sensors are then intersected in order to identify the source location [11].

This constitutes the very simple abstract and geometrical approach to the problem. However, a number of physical phenomena have to be considered in order to make the method reliable. Obviously, the performance of TDOA-based solutions depends very critically on the accuracy and the robustness of the time delay estimation (TDE). One can identify three major problems for TDOA methods for the in-solid case: background noise, reflections (multiple sound propagation paths) and, especially, dispersion. The most crucial problem of in-solid localization is given by the phase velocity dispersion occurring with in-solid wave propagation [12]. The main effect of dispersion is that acoustic waves change their shape in the time domain while they propagate, and therefore the slope of the impulse corresponding to the tactile interaction is modified. The TDOA method is more advantageously used with flat surfaces. However, the same principle can also be applied to curved surfaces, with some additional effort for calculating the hyperbolic curves in the three dimensional space.

2.3 Modular Hardware Platform

In order to create embedded stand-alone applications, that is without using a PC, we are using a modular hardware platform named 'Presto Kit' that we are constantly expanding through the years and that serves as rapid prototyping kit for many of our projects. The modular architecture of the platform is basically composed of four different types of boards that can be combined together:

-        I/O (eg. audio, video, sensors, LEDs, relays)
-        Processing (fixed point, floating point DSP's)
-        Communication (eg. Ethernet, FireWire, USB, Wireless, MIDI)
-        Integrated (microcontroller + I/O + Com.)

Several options exist or are under development for each kind of boards, allowing users to choose the right combination for each application. Boards are stacked one above the other, thanks to a common bus for data transfer that is crossing them vertically (Figure 3).

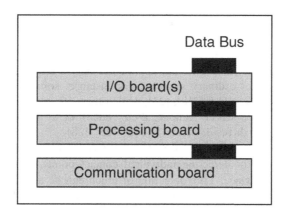

Fig. 3. Modular architecture of the Presto Kit

In those examples with Tangible Acoustic Interfaces, the configuration is composed of a specifically designed preamp board for acoustic sensors (8 channels), a high sampling rate acquisition board (up to 384 KHz), a digital signal processing (DSP) board, and a MIDI board for connecting to a sound module (Percussion Tray), or a relay board for controlling the LED's (Light Globe). During the development phase of algorithms, we were using a FireWire communication board in order to stream signals into Matlab application. Figure 4 shows the first three boards.

Depending on the need, the system can run with uClinux operating system, or with lighter ZottaOS.

Fig. 4. Presto Kit with signal conditioning board for acoustic sensors (bottom), high-speed ADC board (middle) and processing board with BlackFin DSP

3 Computer Vision Approach

Multi-touch screens and surfaces are becoming more and more popular. However, most of the available technologies and approaches only work in specific conditions and are not suitable for ordinary surfaces. For instance, some sensing systems are embedded into the surface [13], [18], [24], while others are specific to screens, either as an overlay above the screen (iPhone, iPod Touch), or as a vision system placed behind a rear projected diffusion screen [4], [5], [23], [28].

Solutions exist to track multiple fingers on a generic surface [14], [15], [22], but they are not suitable for detecting individual contact points, that is, if fingers are touching or not the surface. True detection of touch can be achieved roughly using stereoscopy [16], [21], or more precisely with four cameras placed in the corners of the interactive area [17], [27]. It can also be achieved with a single camera by analyzing the shadow of the fingers [20], or by watching fingers intercepting a plane of infrared light projected above the surface [19]. Virtual Keyboards currently on the market [25], [26] are based on this approach, which has the advantage of requiring less computational power than the other ones. However, those devices do not compute true coordinates of touch and their interactive area is limited to keyboard size. We have adapted this method to be compatible with larger surfaces, and combined it with acoustic onset detection in order to get precise timing information. In addition to fingers, our system can detect oblong objects striking the surface, like sticks and pencils, and it is also suitable to measures the intensity of taps or impacts, allowing to perform the interface both with percussive and touch gestures.

3.1 Multi-touch Detection

At first, a plane of infrared light is created about 1 cm above the surface by using two laser modules equipped with 'line generator' lenses. When fingers or other objects are intersecting the plane, the reflected light is detected by an infrared camera as brighter spots in the image (Figure 5).

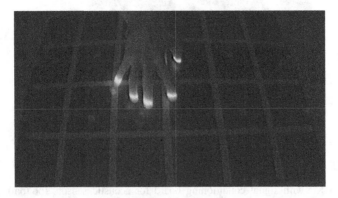

Fig. 5. Image seen by the camera. Visible light is filtered out using a 800nm pass filter.

Simple blob tracking is performed digitally using high-pass filtering, in order to get the finger positions in the image space. Finally, the image positions are converted to the physical space using bi-linear interpolation techniques, after a calibration procedure using a grid of known points.

4 Implementation Examples

4.1 Light Globe

This demonstrator is based on the time reversal technique presented on Chapter 2.1, which relies on the recognition of acoustic signatures. The advantage of this principle is that objects of virtually any shape - flat, non-flat, or irregular - can be transformed into input interfaces with a finite number of interaction points. Usually, one sensor is sufficient, as long as it can be fixed away from a symmetry axis. In this case, any point on the object will have a unique acoustic signature recorded at the sensor level. However, with circular, cylindrical, and spherical objects, there is an infinity of symmetry axis, which means that wherever the sensor is fixed, there will always be two points with the same acoustic response. In order to avoid this, it is necessary to fix two sensors, taking care that both are not on the same symmetry axis. Therefore, in this example, two sensors were used, and placed inside the base of the light globe.

Fig. 6. The light globe with LED's placed under the surface and indicating the position of sensitive spots

As explained in Chapter 2.1, the tactile detection based on Time Reversal is a two-steps procedure, with a training phase, and a running phase. During the training phase (or learning phase), users have to tap with their nail (or another hard object like a ball pen) a few times in order for the system to record the specific acoustic signature. It is thus necessary to have some kind of visual feedback in order to recognize those points. In this demo, the visual feedback is given by a certain number of LED's placed randomly behind the surface of the globe. During the training phase, all LED's are activated. For the running phase, two interactive scenarios were implemented. In the first one, the different spots would act as simple switches. Tapping on a colored spot would simply switch off the corresponding LED. Once all LED's are off, the cycle starts again. For the second scenario, a small game was implemented. LED's are flashing one after the other randomly and users have to 'catch' them by tapping on the active spot before it is turned off again. When successful, the LED remains activated until all others are caught as well. Then a new cycle starts again but faster, that is with less time between two flashing LED's.

4.2 Percussion Tray

The percussion tray demo is based on the TDOA method presented in Chapter 2.2. Four piezoelectric sensors are placed close to the corners of a plastic tray and connected to an electronic device (see section 4.4 for a more detailed description), for calculating the positions of fingers or sticks hitting the tray's bottom (Figure 7).

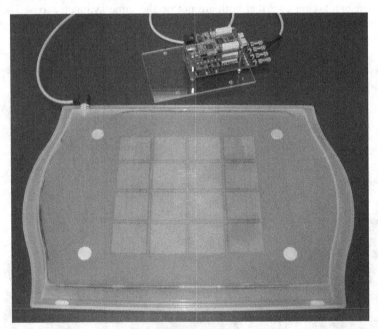

Fig. 7. The Percussion Tray with four acoustic sensors connected to the processing module

In this example, the interface is used as a drum pad. The electronic device is connected via MIDI (Musical Instrument Digital Interface) to an external drum module, where sampled sounds are triggered according to the hit location. For this purpose, a tracing paper with a printed grid of 16 pads is glued on the tray to have a visual reference. The tray is used up side down, and since it is slightly transparent, the tracing paper is visible from the opposite face.

An important feature for such musical application is that there is no noticeable delay in the sonic response when tapping on the tray. Also it is possible to retrigger sounds very quickly, like rolls, up to 20 times per second, with a precision of 1 cm. Previous experiences with tap positions calculated on a personal computer showed it was not possible to reach such performances because of the delay required for signal transmission.

4.3 Multi-touch Table

Figure 8 gives an overview of the setup. The system is comprised of an infrared camera placed above the upper edge of the surface, and two laser modules placed in the corners of the surface one wants to make touch sensitive. The laser modules also contain piezoelectric sensors for sensing the intensity and nature of impacts. They are connected to an electronic board for controlling the power of lasers and processing sensor information. The chosen camera is an OptiTrack Slim:V100 [29], which features embedded blob tracking at 100 fps, allowing for much faster performances and reduced CPU usage than using a normal camera. Also, this camera provides a sync link, which allows for synchronizing the lasers with the shutter of the camera, resulting in an increased signal to noise ratio. Figure 9 shows the system in use, with a simple wood board. The obtained precision is of about 3mm. An interesting possibility, for instance, is to create tool bars or control menus simply by drawing them on a piece of paper and tape them on the surface, like shown on Figure 9.

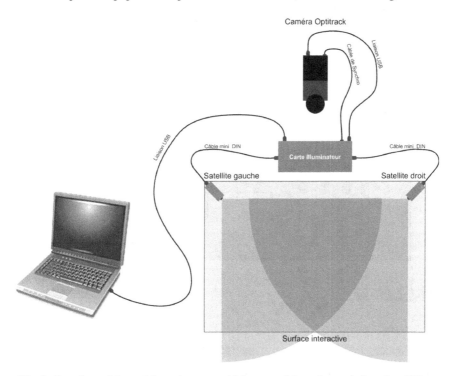

Fig. 8. Overview of the multi-touch setup, with laser modules, electronic board and IR camera

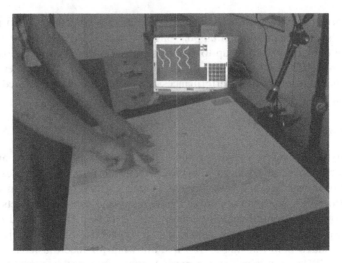

Fig. 9. Simple wood board transformed into multi-touch input interface. A menu with various functions is drawn on a piece of paper on the right side of the table, and used in conjunction with a simple drawing application.

4.4 Surface Editor

The Surface Editor is a graphical software tool that we developed in order to facilitate the creation of interactive scenarios and prototype applications. User interfaces are

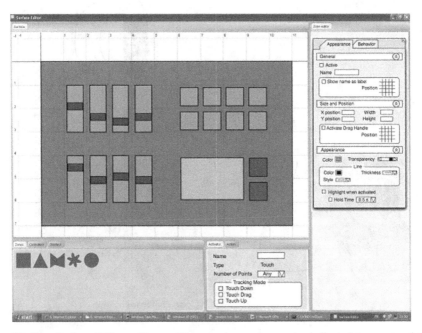

Fig. 10. The Surface Editor with specific mapping components dedicated to create virtual musical controllers

created easily by placing mapping components on the editor's screen, which represents the surface of the interface, and adjusting their parameters. Mapping components are defining the relationship between physical actions (input gestures) and programmed actions (output actions). Each mapping component is implemented in the form of a plug-in and can therefore be custom designed for each particular application. Typical mapping components include GUI elements such as knobs, faders, buttons, etc. The editor is also providing a visual feedback to users, which can optionally be projected on the interface. Figure 10 is showing an application example of the Surface Editor for creating reconfigurable musical controllers. Each component is configured for sending MIDI events to a synthesizer or sound module connected to the computer.

5 Conclusion

We presented several techniques and tools for transforming daily life objects into tactile interfaces. Each of the methods and approaches has their own advantages and disadvantages. Application developers have to chose which one is the best appropriate for a particular application, in particular in relation with the context of use. For instance, acoustic techniques have an advantage in term of integration for public installations, since all sensing part can be hidden behind the surface or inside the object. On the other hand, the computer vision technique presented here allows for more sophisticated interaction, with continuous sensing of multiple touch points, but can be perturbed by direct sun light. No one technique is better than another and that's why we continue to develop both the acoustic and computer vision approaches, in order to face various applications contexts.

It must also be said that most techniques and prototypes presented here were initially developed with the intension to create new musical instruments. The fundamental statement was to be able to use vibrating elements as input interfaces. Therefore, it was necessary to find non-intrusive touch technologies that allow for using existing objects and structures without modifying them. By extension, this allowed for using daily life objects, opening the way for many other applications, like smarts objects and context-aware environments, and more generally for creating more natural and transparent user interfaces. Traditional sensing techniques have the disadvantage of requiring mechanical or electronic devices at the point of interaction with the interface (switches, potentiometers, sensitive layers, force resistive sensors, RFID, etc). This increases manufacturing costs and limits the interaction to predetermined points or surfaces fitted with the appropriate sensing technology. The methods presented here suppress the need for intrusive sensors or devices in the area of contact and interaction. Instead, the entire object becomes like a giant sensor, allowing for more extended interaction areas, and thus providing a possible path towards the creation of 'smart surroundings'.

More information on the recent developments on www.future-instruments.net.

Acknowledgments. The projects presented here have been supported by the FP6 European project TAI-CHI, the Swiss National Funding Agency and the University of Applied Sciences. Special thanks to all the people involved in those developments, in particular Arnaud Guichard for the development of TAI algorithms and their implementation on the DSP platform, Mathieu Kaelin for the development of the hardware part of the multi-touch system, Aymen Yermani for the initial work on the computer vision algorithms, and Greg Kellum for his work on the software tools.

References

[1] Ishii, H., Wisneski, C., Orbanes, J., Chun, B., Paradiso, J.: PingPongPlus: Design of an Athletic-Tangible Interface for Computer-Supported Cooperative Play. In: International Conference on Human Factors in Computing Systems, CHI 1999, Pittsburgh, Pennsylvania, USA, May 15–20 (1999)

[2] Paradiso, J.A., Leo, C.K., Checka, N., Hsiao, K.: Passive Acoustic Sensing for Tracking Knocks Atop Large Interactive Displays. In: Proceedings of the 2002 IEEE International Conference on Sensors, Orlando, Florida, June 11-14, 2002, vol. 1, pp. 521–527 (2002)

[3] Paradiso, J., Leo, C.-K.: Tracking and Characterizing Knocks Atop Large Interactive Displays. Sensor Review (special issue on vibration and impact sensing) 25(2), 134–143 (2005)

[4] Han, J.Y.: Low-Cost Multi-Touch Sensing through Frustrated Total Internal Reflection. In: Proceedings of the 18th Annual ACM Symposium on User Interface Software and Technology (2005)

[5] Jordà, S., Kaltenbrunner, M., Geiger, G., Bencina, R.: "The reacTable*". In: Proceedings of the International Computer Music Conference (ICMC 2005), Barcelona, Spain (2005)

[6] Crevoisier, A., Polotti, P.: Tangible Acoustic Interfaces and their Application for the Design of new Musical Instruments. In: Proceedings of NIME 2005 (2005)

[7] Crevoisier, A., Polotti, P.: A New Musical Interface Using Acoustic Tap Tracking. In: MOSART Workshop on Current Research Directions in Computer Music, Barcelona, November 15-16-17 (2001),
http://www.iua.upf.es/mtg/mosart/papers/p26.pdf

[8] http://www.taichi.cf.ac.uk/

[9] Ing, R.K., Catheline, S., Quieffin, N., Fink, M.: Dynamic focusing using a unique transducer and time reversal process. In: Proceeding of the 8th International Congress on Sound and Vibration, pp. 119–126 (2001)

[10] Ing, R.K., Quieffin, N., Catheline, S., Fink, M.: Tangible interactive interface using acoustic time reversal process. Applied Physics. Lett. (2005)

[11] Rindorf, H.J.: Acoustic Emission Source Location in Theory and Practice. Brüel and Kjær Technical Review Nr 2 - 1981, Nærum Offset, Denmark (1981)

[12] Polotti, P., Sampietro, M., Sarti, A., Tubaro, S., Crevoisier, A.: Acoustic Localization of Tactile Interaction for the Development of Novel Tangible Interfaces. In: The 8th Int. Conference on Digital Audio Effects, DAFX 2005, Madrid, Spain, September 20-22 (submitted, 2005)

[13] Dietz, P.H., Leigh, D.L.: DiamondTouch: A Multi-User Touch Technology. In: Proc. of the ACM Symposium on User Interface Software and Technology (UIST) (2001)

[14] Koike, H., Sato, Y., Kobayashi, Y.: Integrating Paper and Digital Information on EnhancedDesk: a Method for Realtime Finger Tracking on an Augmented Desk System. ACM Transactions on Computer-Human Interaction (TOCHI) 8(4), 307–322

[15] Letessier, J., Berard, F.: Visual Tracking of Bare Fingers for Interactive Surfaces. In: Proc. of the ACM Symposium on User Interface Software and Technology (UIST) (2004)

[16] Malik, S., Laszlo, J.: Visual Touchpad: A Two-Handed Gestural Input Device. In: Proceedings of the International Conference on Multimodal Interfaces, pp. 289–296 (2004)

[17] Martin, D.A., Morrison, G., Sanoy, C., McCharles, R.: Simultaneous Multiple-Input Touch Display. In: Proc. of the UbiComp 2002 Workshop (2002)

[18] Rekimoto, J.: SmartSkin: An Infrastructure for Freehand Manipulation on Interactive Surfaces. In: Proceedings of CHI 2002, pp. 113–120 (2002)

[19] Tomasi, C., Rafii, A., Torunoglu, I.: Full-size Projection Keyboard for Handheld Devices. Communications of the ACM 46(7), 70–75 (2003)

[20] Wilson, A.: PlayAnywhere: A Compact Tabletop Computer Vision System. In: Proceedings of the ACM Symposium on User Interface Software and Technology (UIST) (2005)

[21] Wilson, A.: TouchLight: An Imaging Touch Screen and Display for Gesture-Based Interaction. In: Proceedings of the International Conference on Multimodal Interfaces (2004)

[22] Wu, M., Balakrishnan, R.: Multi-finger and Whole Hand Gestural Interaction Techniques for Multi-User Tabletop Displays. In: Proc. of the ACM Symposium on User Interface Software and Technology (2003)

[23] http://www.surface.com

[24] http://www.tactex.com

[25] http://www.celluon.com

[26] http://www.lumio.com

[27] http://www.smarttech.com

[28] http://nuigroup.com/wiki/Diffused_Illumination_Plans/

[29] http://www.naturalpoint.com/

Using a Movable RFID Antenna to Automatically Determine the Position and Orientation of Objects on a Tabletop

Steve Hinske[1] and Marc Langheinrich[2]

[1] Inst. for Pervasive Computing
ETH Zurich
8092 Zurich, Switzerland
`steve.hinske at inf.ethz.ch`
[2] Faculty of Informatics
University of Lugano (USI)
6904 Lugano, Switzerland
`langheinrich at acm.org`

Abstract. Augmented tabletop games support players by sensing the context of game figures (i.e., position and/or orientation) and then using this information to display additional game information, or to perform game related calculations. In our work we try to detect the position and orientation of game figures using small, unobtrusive passive RFID tags. In order to localize our multi-tagged objects, we use a small movable antenna mounted underneath the table to scan the game environment. While this approach is not capable of real-time positioning, it achieves a very high accuracy on the order of a few millimeters. This article describes our experimental setup, discusses the trade-off between speed and accuracy, and contrasts our approach with a multi-antenna setup.

Keywords: Radio-Frequency Identification (RFID), Object Detection, Miniature War Games, Localization.

1 Introduction

Using radio frequency identification (RFID) technology for detecting tagged objects on surfaces such as shelves or tables has been subject to some research in recent years. Scenarios such as smart shelves in retailing [1,3,10] or tabletop gaming applications [2,7] can greatly benefit from this unobtrusive localization and identification technique. For many applications, it suffices to know whether a given object is in read range of the antenna (i.e., whether the object is *there*), but some require further information such as where an object is exactly and maybe even how it is oriented.

Previous work on using RFID technology for determining the position of objects used multiple antennas organized in a specific way (e.g., a chessboard pattern) to collaboratively infer the exact position and orientation of multi-tagged objects [4,5]. While the reported prototypes were able to achieve an accuracy of a few centimeters,

D. Roggen et al. (Eds.): EuroSSC 2008, LNCS 5279, pp. 14–26, 2008.

tabletop games typically require accuracy on the order of millimeters. In *miniature war games*, for example, two or more players engage in battle with each other, commanding an army of numerous game objects representing combat units. In such settings, measurements need to be as accurate and precise as possible in order to properly determine, for example, the visibility of enemies, or the range and effect of an attacker's weapon.

Usually, players of miniature war games use rulers and goniometers to measure the distances and angles between units and their orientation. Since these tasks can be laborious and time-consuming, our goal is to support the players by automatically capturing this information and providing it to them in an automated and unobtrusive fashion, allowing them focus on social interaction and on the game itself.

Throughout this paper, we will use the example of *Warhammer 40k*, a popular miniature war game that features numerous game units and landscape components scattered over a large table. Players stand around the table, positioning their units and measuring distances between these units, as shown in Fig. 1. As *Warhammer 40k* continuously requires precise information about the location and orientation of all game objects, it proves to be an excellent example application for a precise positioning system: in order to be of any value to the players, the system must be able to localize objects within a few millimeters.

Fig. 1. A typical Warhammer battlefield scene (from [5])

This paper reports on our investigation into the use of passive RFID technology to not only detect objects on a surface, but to also determine their exact position and orientation. This paper is structured as follows: first, we give a brief overview of localization techniques and elaborate on why RFID technology is beneficial for locating objects on a surface. Second, we describe our initial approach which utilizes an antenna grid to localize tagged objects. Based on these findings, we then propose a new infrastructure that uses only one movable antenna and significantly improves the localization accuracy. The paper is concluded with a short summary and discussion of the presented work.

2 The Benefits of Using RFID Technology

In [4,5] we previously discussed in detail the advantages and disadvantages of using RFID technology for determining the position and orientation of objects. We thus give only a brief summary here.

Compared to the technologies presented in Table 1, RFID technology has several advantages. When examining the working principle and features of (high frequency) RFID technology, it becomes apparent that [5]:

- the technology can be hidden and thus works unobtrusively (small tags),
- the objects are almost maintenance-free (except for exchanging damaged RFID tags),
- there is no need to calibrate the equipment,
- each object is uniquely and unambiguously identifiable,
- no line-of-sight is required, and
- costs are low by comparison.

In the context of a miniature war game, this particularly means that the objects can be moved freely on the surface, even if there are decorative elements. The technology is completely disguised (i.e., the RFID tags are invisibly embedded into the objects and the antenna grid is installed under a table), and can thus unobtrusively support the players' actions. Furthermore, the infrastructure is almost maintenance-free, as we do not need to calibrate the technology and the RFID tags do not need to be maintained or replaced. The calculation can be done by a computer with average computational power, which in turn means that the computer employed can be rather small and thus also be integrated in the environment. Additionally, we can scan many figures simultaneously and unambiguously identify them, which is not as easily possible with other localization techniques. Since the antennas induce a three-dimensional field, it is also possible to have game elements on the table that make the game map uneven or even represent "hovering" units (see Fig. 2 left) or taller buildings (e.g., hills, houses, etc.; see Fig. 2 center and right). Finally, RFID technology is comparably inexpensive compared to other technologies.

Table 1. Overview of the disadvantages of localization techniques for objects on tabletops

Technology / Approach	Disadvantages
Ultra-wideband (UWB) (e.g., [13])	• Not precise enough (within tens of centimeters) • Fixed infrastructure[1] • Rather big tags • Tags require batteries • Requires calibration • Expensive
Ultrasound (infrastructure) (e.g., [11])	• Fixed infrastructure • Tags too big • Tags require batteries • Very expensive
Ultrasound (no infrastructure) (e.g., [6])	• Tags too big • Tags require batteries
Load sensing (e.g., [9])	• Cannot identify objects • Not sensitive enough for lightweight objects • Requires totally flat surface
Infrared technology (e.g., [8])	• Requires line-of-sight (cf. Fig 2)
Visual recognition (e.g., [7])	• Cannot identify similar or equally looking objects (cf. Fig. 2) • Requires calibration • Requires computing power
Touch technology(e.g., [12])	• Requires calibration • Very expensive • Detection is based on capacitive coupling (through the human body)

The main disadvantage of RFID technology, however, is that it was simply not meant to be used for localization, not to mention orientation measuring, but solely for identification. And even when used for localization, it is typically limited to offering binary information only: one can only tell whether a specific object is in read range, but not where exactly[2].

---

[1] *Fixed infrastructure* means here that the sensing devices are physically attached to the environment, for example, the walls or the ceiling (i.e., they cannot be (re-)moved easily). This does not hold true for our RFID *infrastructure*, which could be moved (as can, for example, a camera for visual recognition).

[2] There are readers that can either vary the power level or read the signal strength. The former approach is time-consuming while the latter is rather error-prone, due to possible distortion caused by reflections, multipath effects, etc. And, unfortunately, in both cases the readers are very expensive and thus unsuitable for most everyday applications.

Fig. 2. Objects that are not directly on the ground make it difficult for some position recognition techniques to work (left). Landscape elements (centre) do not work with sensing techniques that require a flat surface and/or line-of-sight. Video recognition does not work with many objects that look very similar, i.e., objects that do not have distinctive shapes (right).

3   Previous Work on Using RFID Technology for Localization

We previously reported on the development of an infrastructure that enabled the automatic and relatively precise tracking of the location and orientation of game objects on the playing field [4,5]. Our approach was to increase the number of antennas in order to exploit the information gained from the overlapping read ranges of the antennas. The general principle is shown in Fig. 3. The circles around the antennas symbolize their read range given a specific tag (the read range inter alia varies with the tag model). This modified version of the *cell of origin* approach allowed us to determine the position of a tag as follows: when an antenna reads a given tag, the grid increases an internal counter for each cell (the small white squares in Fig. 3) that is in range of this particular antenna. After completing the read cycles, the tag is most likely in (one of) the cells with the highest counters.

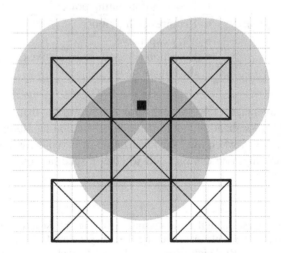

Fig. 3. Multi-antenna approach using multiple antennas organized in a grid (big squares with thick black lines) to determine the position of a tag (small black square) by measuring the overlapping areas of the read ranges (grey circles) (from [4]).

In Fig. 3, the dark area in the center marks the area where the tag, represented by the small black square, must be located. It is not possible to determine where exactly it is within this area. Therefore, the goal is to minimize this area of uncertainty. It is obvious that the size of the "uncertainty area" depends on the number and size of the read range circles (i.e., the antennas), and on the layout of the antenna grid. The smaller the read range circles, the more antennas there are and the denser the grid, the better.

To counter some technical deficiencies of the currently available equipment and the general problem of interference that RFID technology has to cope with (e.g., tags might not be read in a cycle, environmental interference such as metallic objects or even people, etc.), we experimented with several constellations of RFID tags and antennas and varied the following components:

- the layout of how the antennas are placed (design of the antenna grid),
- the RFID antenna model,
- the RFID tag, and
- the read range of the employed reader.

An automated test environment was used to investigate how the variation of these components influences the preciseness of the readings, using two antenna models (FEIG ID ISC.ANT 100/100 and FEIG ID ISC.ANT 40/30), and two different RFID tags. After measuring the range in which each tag could be read by the reader, eight antennas were arranged in a chessboard pattern (see Fig. 4) and a number of objects were tagged with several RFID tags and placed onto the field.

A single reader (FEIG ID ISC.MR 101-A) was connected to the antennas via a multiplexer (FEIG ISC.ANT.MUX 8), which sequentially energizes the individual

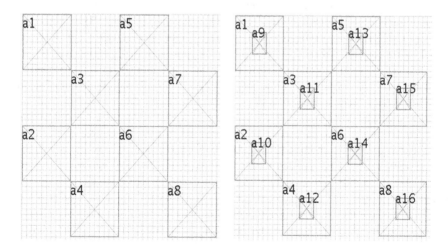

Fig. 4. The two different antenna grid layouts used in [5]. The grid on the left consists of eight 10x10 centimetres antennas, while the one on the right uses eight additional antennas that are considerably smaller (3x4 centimetres) and placed on top of the bigger antennas (from [5]).

antennas to return the read tags in range. After several read cycles (one read cycle took approximately 2-3 seconds) that helped to avoid erroneous read data, our system determined the highest probability for each scanned tag on the board. Based on this data and the known shape and size of each object, the estimated position and orientation of the game figures was then calculated and displayed.

The results as reported in [5] showed that the best estimates of the scanned tags were within a deviation of 3-4 centimeters, rendering this approach insufficient for game applications like miniature war games that require a resolution of less than one centimeter.

4 Improving Resolution Accuracy Using Moveable Antennas

While previous work showed that it is in principle possible to use off-the-shelf RFID technology for determining the position of objects, increasing the accuracy of a multi-antenna approach seems difficult: in order to improve localization results, the antennas would either need to be moved closer together, or smaller antennas would need to be used. Increasing antenna density quickly conflicts with the nature of RF fields: if the antennas are too close together, the field induced by one antenna will be inexorably extended by the coils in the adjacent antennas (i.e., an antenna might then discover a tag that is actually near the coil of another antenna). Using smaller antennas would not only require a lot more antennas (and thus also more readers and multiplexers to power them) to cover an area of the same size, but also limit overall detection rates due to the much smaller read ranges of each individual antenna.

To still benefit from the many advantages of RFID technology and yet compensate the insufficient accuracy, we thus propose a different approach: instead of using many antennas in a grid, we use only one antenna that will be moved across the area. This single antenna will continuously read what RFID tags are in range. This information, combined with the current location of the antenna, can then be used localize the individual RFID tags. In contrast to a multi-antenna approach, this solution has the obvious drawback of requiring more time to cover the same area. However, since tabletop war games do not require real-time positioning[3], a certain period of scanning time migh be perfectly acceptable.

Using a LEGO MindStorms NXT robotic set[4], we constructed a test environment in which a robot controls a slide carriage attached to two orthogonal track systems (see Fig. 5). The RFID antenna (FEIG ID ISC.ANT 40/30) on top of the carriage can then be moved across the surface with an accuracy of a few millimeters. The size of our test bed is ca. 40x50 centimeters (see Fig. 6). The FEIG ID ISC.MR101-A reader antenna can power the antenna every 250ms, which means four read cycles per second. The carriage moves in a zigzag fashion across the board, i.e., it moves along the x-axis to the end of the surface, advances on the y-axis for 1cm, moves back along the x-axis, and so forth.

---

[3] Miniature war games, as most other tabletop games, are turn-based; in fact, one turn can easily take up several minutes.

[4] http://mindstorms.lego.com/

Fig. 5. The robot moves the carriage with the antenna (see white arrow) using two orthogonal track systems (left). The NXT control unit is connected to a computer (right).

Fig. 6. The test environment with the robot, the carriage, and the tracks as well as the cardboard surface with two test objects. The two objects have RFID transponders attached to each of their corners.

Since the carriage is moved by a cogwheel on a track with small cogs (see Fig. 5), it is not possible to indicate the speed in a distance-time relation (e.g., in cm/s) but only in degrees determined by the rotation of the motor. The motor could be controlled in ten steps (10% to 100% of the max. power output), resulting in different velocities of the carriage (see Fig. 7), which also influences the number of possible measurements, since we can energize the antenna at a rate of 4Hz (see Fig. 8).

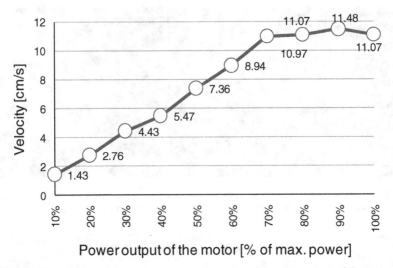

Fig. 7. The velocity of the carriage in dependence on the power output of the motor

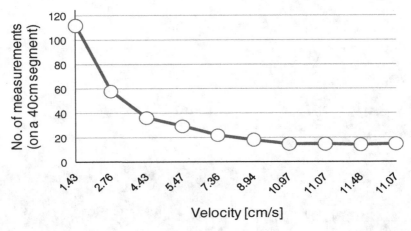

Fig. 8. The number of possible measurements in relation to the velocity

We began our test series by placing two RFID transponders at designated X-Y lo-cations on the test surface: they were positioned at 20cm/10cm and 30cm/30cm, respectively, with the lower left corner being the point of origin. Starting from the zero-point, the carriage would then run over the whole area with a velocity of ca. 4.4cm/s (this initial value was chosen to guarantee approximately one read cycle per cm). We ran the test series five times. Tables 2 and 3 summarize the results.

It is worth noticing that the deviation on the y-axis was constant in each case, but there is a simple explanation for this: while the antenna constantly moves along the x-axis (which elucidates the varying x-values), the y-axis remains fix und thus the deviation constant. In general, with average deviation being between 0.1 and 0.2 centimeters, the measurements were extremely accurate, exceeding our expectations by far: our system

Table 2. Results for the 20/10 coordinates. All values are in centimeter.

	Absolute value of measurement		Deviation (unsigned)	
	x	y	x	y
1	20.1	9.8	0.1	0.2
2	20.0	9.8	0.0	0.2
3	20.0	9.8	0.0	0.2
4	20.1	9.8	0.1	0.2
5	20.1	9.8	0.1	0.2

Table 3. Results for the 30/30. All values are in centimeters.

	Absolute value of measurement		Deviation (unsigned)	
	x	y	x	y
1	30.0	29.9	0.0	0.1
2	30.1	29.9	0.1	0.1
3	30.0	29.9	0.0	0.1
4	30.0	29.9	0.0	0.1
5	30.2	29.9	0.2	0.1

could in most cases determine where a transponder is located with a precision on the order of a few millimeters.

Unfortunately, regardless of this almost perfect accuracy, there was a severe downside: each test round took approx. 11 minutes (ca. 13 seconds for scanning the x-axis incl. adjustment time multiplied by 50 rows), which even without real-time requirements might render this approach infeasible for most practical applications. We therefore intended to considerably shorten the time it takes to scan the surface area, though we knew that increasing the speed of the carriage would lower accuracy due to fewer measurements (see Fig. 8). The aim was to determine the trade-off between the costs (i.e., the loss of accuracy) and the benefits (i.e., shorter read times). To this end, we repeated the previous test series with different velocities and calculated the mean deviation for each velocity. The results are summarized in Fig. 9.

In general, slower speeds show a lower average positioning error. The minor exceptions at velocities of 2.76cm/s and 4.43cm/s seem to result from natural oscillation of the motor when rotating with lower frequencies (the loss of accuracy is also only about 1 millimeter). However, the error increases with higher speeds and reaches some 5 millimeters at velocities of around 11cm/s. The boundary seems to lie at a velocity of 7.36cm/s, which corresponds to a power output of 50%. In this case, scanning an area the size of our test surface (40x50cm) took approximately 7.5min Apparently, it is not possible to further reduce this time with our test setup.

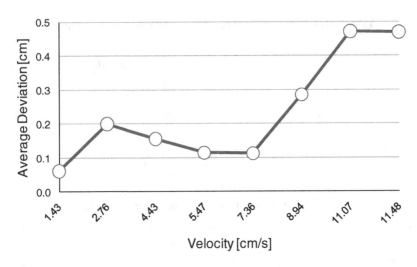

Fig. 9. The average deviation against the velocity

5 Conclusions

In this paper we presented a novel approach to determine the position and orientation of objects using RFID technology. Based on the example of a miniature war game, we demonstrated the idea of using an antenna that is attached to a slide carriage which moves the antenna over (or rather under) a surface with tagged objects. Calculating the position of each single tag using the overlapping read ranges at each reading position of the antenna allows us to estimate where the object is approximately positioned, and thus, in the case of multi-tagging, how it is oriented. As far as we know, there is currently no other sensing infrastructure that is capable of such high accuracy (except for visual recognition, which fails at differentiating equally looking objects though). In principle, it is possible to cover even very large tables by simply sectioning it and applying one scanning set-up to each section.

The downside clearly is the time required for scanning an area. The main objective for future work would thus be to significantly diminish the scan time. Currently, we see two options to achieve this:

- First, we could use a slightly modified version that minimizes the idle read time in which the antenna does not read any transponders (i.e., the time between the reading of one tag and the next one). Consequently, we have to increase the velocity in between transponders, which can be done as follows: we move the carriage with full speed (ca. 11.5cm/s) along every fourth x-axis, omitting three rows, and, if we read at least one tag on the way, we move over the same section again at a slower velocity. Preliminary tests revealed that the time could be significantly reduced to approx. 4.5min while the average deviation was only slightly worse. This approach, however, very much depends on the on the scarcity and distribution of the objects on the surface: with a high number and density of objects, this approach will presumably take longer than with the "naive" version. More tests will need to be conducted to evaluate this properly.

- Second, we could simply split the area into smaller sections and use one antenna for each section. If the sections are, for example, 20x20cm, the time required could be reduced to approx. one minute, which is undoubtedly fast enough for most turn-based games and many other applications (e.g., a smart shelf in a store). In addition to that, we would still need considerably less antennas compared to the antenna-grid approach.

Nonetheless, this proposed localization technique cannot be used for real-time applications, in contrast to other technologies (e.g., visual recognition or ultrasound). Our approach is furthermore subject to criticism in three more points. First, the selection of the individual components (RFID readers, antennas, and tags) is very crucial: if we substitute only one component, the results are at least distorted, if not totally different, requiring re-calibration. Second, RFID does not work too well with metallic objects and environments: metal biases the read rate of readers and tags. Third, we will also need to verify the influence of other game objects on item detection, e.g., houses or hills that figures are placed in or on, which increase the distance between the surface and the tag.

These disadvantages notwithstanding, it seems that this approach based on RFID technology is a promising candidate for localizing objects on a surface: the high accuracy of about 1 millimeter, combined with the inherent advantages of RFID technology (e.g., a completely invisible and unobtrusive scanning infrastructure that also allows for uniquely and unambiguously identifying all objects) promises cheap high resolution localization for augmented tabletop games in the near future. And though augmented tabletop games admittedly are a prime example, one can envision other domains that could benefit from such a high resolution, e.g., architectural and city planning applications that involve 3-dimensional models arranged on a tabletop.

Acknowledgments. The authors would like to thank Thomas Lohmüller for the implementation of the prototypical demonstrator.

References

1. Decker, C., Kubach, U., Beigl, M.: Revealing the Retail Black Box by Interaction Sensing. In: 23rd International Conference on Distributed Computing Systems Workshops (ICDCSW 2003) (2003)
2. Eriksson, D., Peitz, J., Björk, S.: Enhancing Board Games with Electronics. In: PerGames 2005, 2nd International Workshop on Pervasive Gaming Applications at Pervasive 2005 (2005)
3. Fleisch, E., Mattern, F. (eds.): Das Internet der Dinge - Ubiquitous Computing und RFID in der Praxis. Springer, Heidelberg (2005)
4. Hinske, S.: Determining the Position and Orientation of Multi-Tagged Objects Using RFID Technology. In: PerTec Workshop 2007 at IEEE International Conference on Pervasive Computing and Communications (PerCom) (2007)

5. Hinske, S., Langheinrich, M.: An RFID-based Infrastructure for Automatically Determining the Position and Orientation of Game Objects in Tabletop Games. In: Magerkurth, C., Röcker, C. (eds.) Concepts and Technologies for Pervasive Games - A Reader for Pervasive Gaming Research, vol. 1, pp. 311–336. Shaker Verlag (2007)
6. Krohn, A., Zimmer, T., Beigl, M.: Enhancing Tabletop Games with Relative Positioning Technology. In: Advances in Pervasive Computing, Österreichische Computer Gesellschaft (2004)
7. Magerkurth, C., Stenzel, R., Prante, T.: STARS - A Ubiquitous Computing Platform for Computer Augmented Tabletop Games. In: Ljungstrand, P., Brotherton, J. (eds.) UbiComp 2003. Springer, Heidelberg (2003)
8. Mandryk, R.L., Maranan, D.S., Inkpen, K.M.: False Prophets: Exploring Hybrid Board/Video Games. In: Extended Abstracts of CHI 2002 (2002)
9. Schmidt, A., Van Laerhoven, K., Strohbach, M., Friday, A., Gellersen, H.-W.: Context Acquistion based on Load Sensing. In: Borriello, G., Holmquist, L.E. (eds.) UbiComp 2002. LNCS, vol. 2498. Springer, Heidelberg (2002)
10. TecO, Ubicomp Research and Projects at TecO: SmartShelf,
    http://www.teco.unikarlsruhe.de/research/ubicomp/smartshelf/
11. The Bat System, 3D Ultrasonic Positioning for People and Objects, University of Cambridge,
    http://www.cl.cam.ac.uk/Research/DTG/research/wiki/BatSystem
12. Tse, E., Greenberg, S., Shen, C., Forlines, C.: Multimodal Multiplayer Tabletop Gaming. In: PerGames 2006, 3rd International Workshop on Pervasive Gaming Applications at Pervasive 2006, pp. 139–148 (2006)
13. Ubisense System, http://www.ubisense.net

Vision-Based Detection of Mobile Smart Objects

David Molyneaux[1], Hans Gellersen[1], and Bernt Schiele[2]

[1] Computing Department, Lancaster University, England
{d.molyneaux,hwg}@comp.lancs.ac.uk
[2] Computer Science Department, Darmstadt University of Technology
schiele@informatik.tu-darmstadt.de

Abstract. We evaluate an approach for mobile smart objects to cooperate with projector-camera systems to achieve interactive projected displays on their surfaces without changing their appearance or function. Smart objects describe their appearance directly to the projector-camera system, enabling vision-based detection based on their natural appearance. This detection is a significant challenge, as objects differ in appearance and appear at varying distances and orientations with respect to a tracking camera. We investigate four detection approaches representing different appearance cues and contribute three experimental studies analysing the impact on detection performance, firstly of scale and rotation, secondly the combination of multiple appearance cues and thirdly the use of context information from the smart object. We find that the training of appearance descriptions must coincide with the scale and orientations providing the best detection performance, that multiple cues provide a clear performance gain over a single cue and that context sensing masks distractions and clutter, further improving detection performance.

Keywords: Cooperative Augmentation, Smart Objects, Vision-Based Detection, Natural Appearance, Multi-Cue Detection.

1 Introduction

Smart objects research explores embedding sensing and computing into everyday objects – augmenting objects to be a source of information, for example, on their identity, state, and context in the physical world [1]. One challenge for the design of such smart objects is to preserve their original appearance and function, exploiting a user's familiarity with the object. However, we often want to augment objects with digital state information or reveal knowledge hidden inside the object, as this allows users to address tasks in physical space and receive direct visual feedback from the objects themselves. Recent work has proposed cooperation with projector-camera systems in the environment [2], allowing objects to be augmented with interactive projected displays on their surfaces without embedding screens or permanently changing their appearance. The ability for objects to function as both input and output medium simultaneously enables scenarios such as objects that monitor their physical condition and visually display warnings if these are critical [1].

A central problem for realisation of displays on everyday objects with the Cooperative Augmentation concept proposed in [2] is object detection and tracking with

D. Roggen et al. (Eds.): EuroSSC 2008, LNCS 5279, pp. 27–40, 2008.
© Springer-Verlag Berlin Heidelberg 2008

vision-based techniques. Only when an object is detected can the projector-camera system align its projection so the image is registered with the object's surfaces. In experimental prototypes detection is commonly achieved with fiducial markers [3]. However, with a view to ubiquitous augmentation of objects, it is more realistic to base detection on the natural appearance of objects. This detection is a significant challenge in real-world environments, as objects naturally vary in their appearance – hence we assume that multiple methods should be provided as alternatives for detection. A second challenge that arises from realistic scenarios is that objects will appear at varying distances and orientations with respect to a tracking camera as they are picked up, moved around and manipulated by users.

Instead of storing the knowledge required to detect and track objects in a smart environment, the Cooperative Augmentation concept [2] distributes the knowledge and intelligence into the smart objects themselves. Hence enabling projector-camera systems to become ubiquitous, as they merely offer a generic projection service to all smart objects. The smart objects store an "Object Model" containing knowledge of their appearance (in terms of numerical appearance description data extracted from training images of the object by computer vision detection algorithms), a 3D model and sensors, and actively control the interaction with projector-camera systems.

Objects cooperate by describing their appearance to the projector-camera system on entry to the environment. The projector-camera system supports flexible amounts of appearance knowledge and differing object appearances by implementing multiple detection methods in the controlling computer. The detection methods used are dynamically configured in response to the knowledge contained in the object appearance description and the object's context. Examples of context information we can incorporate in detection are: knowledge of whether the object is moving from embedded sensors (in which case we could choose the fastest detection method) and the object's background (e.g. we would not use a colour detection method to detect a blue object in a blue room, as the probability of detection is low due to distraction).

A learning process in the projector-camera system extracts more appearance knowledge about objects over time and re-embeds this back into the smart object for increased robustness in the detection process. This process allows flexible deployment of new detection methods in projector-camera systems, as the learning process will automatically update an object's knowledge to include the new appearance description following initial detection.

In this paper we investigate four vision-based methods for detecting smart objects, representing the different natural appearance cues of colour, texture, shape and features of objects. The contributions of this paper are three experimental studies. The first is targeted to understand the impact of object scale and rotation on different detection methods. As a result of the study, we gain insight into training requirements of different detection approaches and understanding of when to extract new appearance knowledge in the learning process to maximise detection performance.

The second explores how the different cues can be combined to increase detection performance and the third investigates how the use of context information supplied by the smart object can be used to constrain the multi-cue detection process. The results show a detection performance gain observed when multiple appearance cues are combined and a very clear additional performance gain with the use of context information from movement sensing embedded in the target object.

2 Related Work

The question of how to detect objects with projector-camera systems [2, 3, 4, 5, 6] has been investigated by many. For example, Ehnes et al. detected objects visually with fiducial markers in [3]. Borkowski et al. tracked a screen visually by its black border in [5]. Bandyopadhyay et al. track and project on objects using magnetic and infra-red tracking systems in [4]. These systems rely on adding hardware or modifying the appearance of an object to enable detection. In contrast, we detect smart objects by their natural appearance. The natural appearance has to be matched to a model of the object built from training images, despite changing viewpoint and illumination.

Many approaches have looked at detecting an object by natural appearance using a single cue, such as colour, texture, shape and feature-based detection. **Colour** is a powerful cue for humans, for example, in traffic lights or warning signs. In computer vision, Swain and Ballard's colour histograms [7] are shown to be invariant to rotation and robust to appearance changes such as viewpoint changes, scale, partial occlusion and even shape. However, colour histograms are sensitive to changes in light intensity and colour. Histogram matching techniques such as the intersection or chi-square (χ^2) measurement are generally robust, as the histogram representation uses the entire appearance of the object rather than just a small number of interest points.

Texture on an object's surface allows a wider range of techniques to be employed. Template matching approaches using cross-correlation are fast, but susceptible to failure with occlusion and must be trained with varying illumination for robustness [8]. The histogram approach was generalised to multidimensional histograms of receptive fields by Schiele and Crowley to detect texture [9]. The histograms encode a statistical representation of the appearance of objects based on vectors of joint statistics of local neighbourhood operators such as intensity gradient magnitude and the Laplacian operator (Mag-Lap). Experimental results show histograms are robust to partial occlusion of the object and are able to recognise multiple objects in cluttered scenes in real-time using a probabilistic local-appearance hashing approach.

Shape can be described using either a global or local shape description. Typical approaches to global shape use Principal Component Analysis (PCA) methods [10, 11]. In contrast, local shape can be detected using methods that describe the silhouette contours of an object [12, 13], and these can be matched to database of object appearances with the object in different poses. For smart objects with known 3D models this database can be created directly by rendering the model in different poses and extracting the silhouette contour using edge detection. The Shape Context method by Belongie et al. [12] has achieved success in character recognition. The contours of an object are extracted and a set of points detected on them. For each point the algorithm extracts the relative spatial distribution of the remaining points on the shape. A histogram of the number of contour points in each "bin" of the log-polar coordinates can be created to describe each point. Histograms are similar for corresponding points; hence objects can be detected by histogram matching.

Local feature based detection algorithms aim to uniquely describe (and hence detect) an object using just a few key points. By extracting a set of interest points such as corners or blobs from training images of an object we can use the local image area immediately surrounding each point to calculate a feature descriptor, which we assume serves to uniquely describe and identify the point. A database of local features registered to a 3D model of the object is constructed off-line. At runtime interest points are detected in the camera image. Object detection now becomes a problem of matching features between the training set and camera image by comparing feature descriptors. One of the most widely used local feature algorithms is Lowe's Scale Invariant Feature Transform (SIFT) [14], however, readers are referred to [15, 16] for an in-depth comparison of different feature detection and descriptor algorithms.

Multiple Cues. As objects vary significantly in their appearance, detection based on a single cue such as just colour or shape can perform poorly in the real-world, as no single cue is both general enough and also robust enough to cope with all possible combinations of object appearance and environment. In theory, the combination of complementary cues leads to an enlarged working domain, while the combination of redundant cues leads to an increased reliability in detection [17]. Popular cue combinations are: colour and edges [18], colour and texture [19], intensity, shape and colour [17], shape, texture and depth [20]. However, these cues are fixed at runtime.

Typically, multiple detection cues are fused in particle filter tracking frameworks, allowing multiple target hypotheses to exist simultaneously [17, 18, 19, 20]. In this case the detection of each cue is often assumed to be independent and integrated to contribute to the overall measurement density. Hence, when multiple cues are fused simultaneously, the failure of one cue will not affect the others [18]. Molyneaux et al. propose a different approach using multiple cues to detect smart objects in [2], but where the cues are chosen at runtime based on the appearance knowledge an object contains and the object's context (from embedded sensors or the object background).

3 Vision-Based Object Detection Methods and Evaluation Dataset

We implement and evaluate 4 detection algorithms in the projector-camera system, from those discussed in related work. These correspond to the four object appearance cues of colour, texture, shape and local features.

For colour we use 3D CIE L*a*b* colour histograms [7] as this colour model was empirically found to detect light and dark objects better than Hue-Saturation-Lightness or RGB. Each dimension has 16 bins, 16 values wide. For texture we use rotation invariant 2D Gradient Magnitude and Laplacian multi-dimensional histograms of receptive fields [9] with 32 bins, each 8 values wide. Scale invariance is achieved using a scale space with Gaussian smoothing. We train with $\sigma=2.0$ then use 3 scales in detection, equal to 0.5σ, σ, 2σ. For shape we match 100 points on image contours with Shape Context [12], using 5 radial bin and 12 angle bin histograms. Histograms are made scale invariant by resizing the diameter of the radial bins equal to the mean distance between all point pairs and rotation invariant by averaging the angle of all point pairs and calculating relative angles. For local features we use Lowe's scale and rotation invariant SIFT algorithm [14], with 3 scales per octave and $\sigma = 1.6$. The detection algorithms range from low to high complexity respectively.

We use ten objects as the dataset for experiments, reflecting everyday objects of varying shape, size and appearance (see Figure 1). The largest object was the chair (90x42x41cm), the smallest was the mug (10.5x10.3x7.2cm).

Fig. 1. Experiment Objects (left to right top to bottom): A book, a cereal box, a chemical container barrel, a chair, a card, a football, a mug, a notepad, a product box and a toaster

3.1 Object Appearance Library

The library comprises images of the objects against plain backgrounds, for training the algorithms. Images were acquired with a colour Pixelink A742 camera (1280x1024 pixels) and 12mm lens. For varying scale, images were captured in 5cm intervals between 1 and 6m from a horizontal fixed camera (10x100=1000 images), as shown in Figure 3. For rotation, images of the objects were captured at 2m, 3m, 4m and 5m from a fixed camera. At each distance the objects were rotated in 10° intervals for a full 360° around the vertical axis with a turntable (4x36x10=1440 images total). The camera was 1.5m above the turntable, with 40° declination. All images were manually annotated with a bounding box for ground truth object location.

3.2 Video Test Library

Synchronised sensor data and video of each smart object moving in a cluttered lab environment was captured at 10fps for 20s from a fixed camera location 2.15m high, with 25° declination. The objects were handheld and moved at a constant walking pace (approximately 0.75m/s) in a 15m path shaped like a 'P' from first entry through a door 5m from the camera, as seen in Figure 2. The object moved around the loop of the 'P' towards the camera (the tip is 2m from the camera), then returning to the door. The test videos include challenging detection conditions such as scaling, rotation around the object's vertical axis and motion blur. Distractions were also present (from other objects or areas of the scene with similar appearances) and the limited camera field of view caused partial occlusion in some frames, hence the videos reflect realistic detection scenarios. Each captured frame was annotated with a bounding box for ground truth object location.

Fig 2. Video images of (left) chemical container and (right) cereal box, including examples of object scaling, rotation, distracting objects, motion blur and partial occlusion

A Particle Smart-Its device in the object sent data from an IEE FSR152 force sensor wirelessly at 13ms intervals and was abstracted to simple object moving or not-moving events by using a threshold on the mean value calculated over a window of 20 samples (260ms). This was set empirically so the object generated continuous "non-moving" events when placed on a surface and "moving" events when mobile.

4 Scale and Rotation Experiments

One of the challenges for visual detection algorithms is to reliably detect objects. Real-world objects are composed of different structures at different scales, which means in practice they appear different depending on the scale of observation. This effect can be seen in Figure 3 (left), where different numbers of corners are detected and in different locations when the cereal box is at different distances to the camera.

When detecting an object in an unknown scene, there is no way to know at which scale the object will appear, as the distance to the object is unknown. In theory, by representing the object or camera image at multiple scales in a scale-space, detection algorithms can be made scale-invariant. A scale-space is built by successively smoothing an image using a Gaussian kernel with an increasing standard deviation (σ), to remove more and more fine detail [21]. We can now try to match the object at different scales or choose the most appropriate scale.

Fig 3. (Left) Corner features (yellow dots) are detected and in different amounts and locations on the Cereal box object at 1m, 3m and 6m distance with the single-scale Harris algorithm [22], (right) Camera and Object Coordinate System Transformations

As seen in Figure 3 (right), rotation of an object can be decomposed into 2D rotation in the camera plane r_z, (equivalent to rolling the camera) and general 3D rotation r_y, r_x (equivalent to changing the camera viewpoint). Detection algorithms can be made invariant to 2D rotation, however, general 3D rotation of an object presents a problem. This is composed of two separate aspects: 1) The appearance of an object surface becomes distorted when it is rotated from being perpendicular to the camera vector t_z. 2) As the object rotates, object surfaces disappear from view and new surfaces come into view.

Our experiments are driven by the three following research questions:

R1) Is scale and 3D rotation an issue for the 4 detection methods we choose?
R2) At what distance do we need to train our 4 methods?
R3) Are some of the 4 methods more robust to scale and rotation than others?

4.1 Design

We perform two series of experiments. The first experiment set investigates scale-invariance in detection algorithms and aims to quantify in what scale range we can repeatably detect an object. All of the algorithms were trained with an image at meter intervals between 1 and 6m (resulting in 6 sub-experiments) and the remaining images of the object appearance image library (99 images) were used for testing. We performed 6 sub-experiments, one for each training distance. The results for all 4 cues are averaged over all objects to give the percentage of all detected objects over the scale range (every 5cm).

The second set of experiments investigates rotation for all 4 methods. For each object we trained the algorithms using a single 0° image from the object appearance library. The remainder of the images between -80° (anti-clockwise rotation of object from 0°) and +80° (clockwise rotation) were used to evaluate the percentage of objects detected with each algorithm. We both trained and tested the algorithms with images of the objects at 3m distance, as this is the centre of our working range.

4.2 Apparatus

A 3.4GHz dual core Pentium-4 running Windows XP SP2 controlled the projector-camera system and was used for all experiments. Algorithms were implemented in C++ using Intel OpenCV API. All experiments in this paper used identical apparatus.

4.3 Procedure

For colour histogram detection the object histogram is back-projected into the camera image and the bounding box of the largest blob above a detection threshold is calculated, representing the most likely object location. For texture detection we use an exhaustive search method, dividing each scale image into a grid of scale-adapted 2D windows of uniform size, with 25% partial overlap between each window. Each window's histogram is calculated and matched against the object description. For shape detection we use a similar approach to texture, but only at a single scale, due to the scale-invariance of the shape context algorithm. Correct detection was assumed for these three algorithms when the detection bounding box had <50% overlap error with the ground truth bounding box from the test image library. For SIFT local features correct detection was assumed when a minimum of 8 features were matched to the training image features using nearest-neighbour Euclidean distance matching and >50% of feature correspondences were correct. Correct correspondences were established based on a manually annotated ground truth homography transformation between the test image and the training image of the object.

4.4 Results

We first present the detection results for scale over all objects, followed by the detection results for 3D rotation. The scale graphs show three training distances – the two extremes of 1m, 6m and the best performing training distance in-between.

Figure 4 (left) and (right) shows detection performance over all objects for SIFT local features and for Shape Context respectively. For SIFT it is clearly visible that the percentage of objects detected falls below 50% after 4.5m distance when trained at 1m. For the chosen scale range, the average percentage of objects detected is highest when trained at 2m (M=82.31, SD=14.34). However, the detection percentage varies least across the scale range when trained at 6m (M=74.44, SD=9.32). For

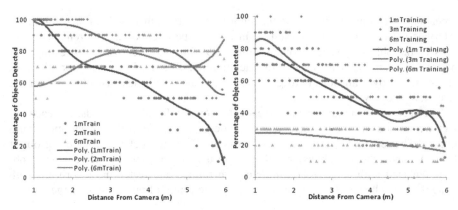

Fig 4. (Left) Detection performance over all objects for SIFT Local features at 1m, 2m and 6m training distances, (right) Performance over all objects for Shape at 1m, 3m and 6m distances

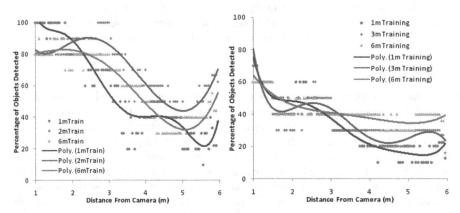

Fig 5. (Left) Detection performance over all objects for Texture at 1m, 2m and 6m training distances, (right) Performance over all objects for Colour at 1m, 3m and 6m training distances

Shape Context the detection percentage is lowest when we train at 6m, never increasing above 30% across the scale range. When trained at 1m and 3m, the results show a downward trend with similar detection performance, however, the highest percentage of objects were detected at when trained 1m (M=56.10, SD=19.71).

Figure 5 (left) and (right) shows detection performance over all objects for Texture and Colour respectively. The results for both experiments show an overall downward tendency for all training distances, with little difference in performance between the training distances. For Texture, the highest percentage of objects were detected with a training distance of 2m (M=72.74, SD=26.59). Here, 50% or more of the objects are detected between 1m and 6m, with the exception of 4.5m to 5.5m where, unusually, the performance drops. In contrast, for Colour the highest percentage of objects are detected when we train at 6m (M=38.05, SD=46.81). The standard deviation for all colour training distances is high, indicating a high variability of detection performance between objects. In fact, 50% of the objects are not detected by colour.

Figure 6 (left) shows detection repeatability over all objects at 3m, for SIFT local features with rotation. The curve is bell-shaped and shows a sharp fall off as we rotate objects away from 0°. Around 20° rotation the repeatability falls to around 40% on average, for all objects. This means only 40% of the original training image features are still being detected. This performance is object dependant. For example, the book object has repeatability greater than 40% between -40° and 40°, peaking at 10° (65.25%). In contrast, the barrel's repeatability is only ever above 40% at 10°, where it peaks (42.11%). Figure 6 (right) shows the percentage of objects detected at 3m, for the remaining three algorithms. For both Texture and Shape the curves have a bell-shaped trend, similar to SIFT. The Colour algorithm varies between 80% and 50% of objects detected, but does not show a decreasing tendency with increasing rotation.

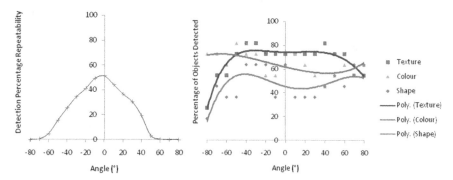

Fig 6. (Left) Detection repeatability of SIFT Local features over all objects when varying 3D rotation angle, (right) Detection performance over all objects when varying 3D rotation angle, for Colour, Texture and Shape; both at 3m training and test distances

4.5 Discussion

The results from the experiments to address R1 to R3 show it is important to consider scale and rotation effects when designing a detection system with multiple cues. For example, for scale, we learned that different algorithms perform best when trained with the object at different distances to the camera. Shape performs best when trained at 1m, Local Features and Texture are best at 2m and Colour performs best at 6m. These distances are the optimum distances for the learning process to extract new appearance information for the respective algorithms, as they will provide the highest overall detection performance throughout the scale range.

Similarly, the algorithms perform differently when we rotate objects. Here, colour does not exhibit a strong decreasing tendency with rotation, suggesting that one view-point may be enough to detect a uniformly colourful object in any pose, and only a small number of viewpoints are required for 3D objects with non-uniform surfaces (6 are recommended in [7]). In contrast, the more bell-shaped curves for the other methods indicate that to detect an object in any pose we need to extract information from many more viewpoints.

The fact that detection performance results when scaling and rotating objects were often not close to 100% around the distance or angle where we train our detection system also suggests that some of the objects were not detected by a particular method or there is a large inter-object performance variability with the algorithms. These results support both our starting assumption that different detection methods behave differently and hence the argument for using multiple detection methods.

5   Multi-cue Detection Experiment

Having found that different detection methods behave differently in section 4, we propose the following research question:
R4) How does the use of multiple-cues change overall visual detection performance?

5.1   Design

In order to perform the experiment, we first trained all 4 algorithms. For each object, then for each detection algorithm, an appearance description was trained using the rotation images in the object appearance library. Appearance descriptions for all detection algorithms were trained using the object ground truth bounding boxes and images of the object at 3m. 6 viewpoints were trained in the horizontal plane, in rotation intervals of 60° around the object's vertical axis.

5.2   Procedure

The 4 algorithms ran on the video test library for all objects. The apparatus and procedure for obtaining the detection result and determining correct detections was identical to the previous experiment. Detection performance results were obtained for each cue in terms of the percentage of video frames with correct detections. These were combined to produce results for multiple-cues by assuming detection is correct in a frame when any of the 4 cue hypotheses correctly detects the object

5.3   Results

Each detection cue was ranked by its detection performance for each object. The overall detection performance achieved by multi-cue combination, together with the breakdown of cue contribution to this performance figure (shown as column colour) can be seen in Figure 7 (left). It is clearly visible that by using a combination of the highest performing cue (lowest colour of the column) together with other cues, the overall detection performance improves for all objects in the study.

The Local Features algorithm proved to be the best single cue for 70% of the objects. Over all objects, the mean average performance increases from 37.89% (with just the highest performing single cue) to 43.47% with all 4 cues, an improvement of 4.96%. The largest individual performance increase is for the book, which improves from 71.52% to 86.71% (a 15.19% increase in the number of frames detected) with the addition of a second cue. The smallest increase is for the Card object (0.62%). The most frequent combination of two highest performing cues was Local Features and Shape, ranked as the best combination for 50% of all objects.

Addition of a third cue produces only a small performance increase of 0.62% over the performance obtained with two cues. The largest increase is an additional 2.47% of detection performance for the ball, while the book, box, cereal box, chair, mug and toaster did not increase in performance by adding a third cue. The addition of a fourth cue did not improve detection performance in any object.

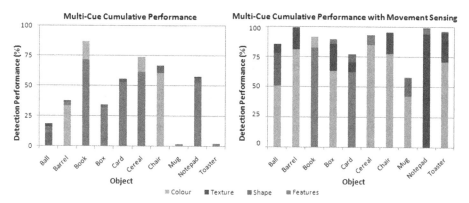

Fig. 7. (Left) Multi-cue cumulative detection performance using 4 cues, (right) Multi-cue performance using 4 cues and context information from movement sensing

5.4 Discussion

The performance benefits from using multi-cue detection are shown in Figure 7 (left). All objects receive an increase in detection performance by using the multiple cues for detection; however, the performance increase averaged over all objects was relatively small (4.96%). This improvement is primarily produced by combining the two best ranked performing single cues. Increasing the number of cues beyond the best two produces only a very small performance increase (0.62%) for the additional computational cost. With 4 cues the objects are detected in only 43.47% of frames.

The cue rankings changed with the objects and each of the cues we studied was best for at least of one of the objects (excluding texture). This again supports the observation from the scale and rotation experiments, that there is large inter-object performance variability with the cues, hence different objects require different detection methods and we can infer that the cues we use are complimentary.

The multi-cue results presented are a best-case scenario, where appearance knowledge is available for all cues and the ranking of cues for each object is known a-priori. We may not initially know the best ranking of cues in new objects; however, in practice the Cooperative Augmentation system can maintain detection metrics such as detection performance and algorithm runtime for each method and use these as additional knowledge during detection method selection. This valuable knowledge can be embedded into the smart object so it is not lost on object departure.

Unless multiple cues can be run in parallel, multi-cue detection is also a trade-off, as each additional detection cue causes a corresponding increase in run-time per camera frame. Consequently, the detection performance increase must be balanced against the processing capabilities in the projector-camera system.

Storing multiple appearance descriptions for multi-cue detection also has a large overhead in terms of memory storage and battery cost for transmission from the sensor node to the projector-camera system. Consequently, in cases where Object Models are too big or transmission is too costly, we assume they are stored in a network resource and only the corresponding URL is embedded in the smart object.

6 Multi-cue Detection with Context Information Experiment

As shown in the previous experiment, while the use of multiple cues has benefits, still only a low overall detection performance is achieved. The detection results are also lower than the results reported for the algorithms in [7, 9, 12, 14], however, these results were for objects on plain backgrounds or with dissimilar distracting objects. Hence, we hypothesise that the "realistic" environment with clutter and distraction in the videos is causing poor detection results, and performance can be increased by making detection robust to this. We therefore propose the research question:

R5) How does the addition of context information constrain the detection process?

6.1 Procedure

The experiment using context in detection used an identical design and procedure as the multi-cue experiment. However, now the absolute difference between the current and previous video image was calculated to obtain a basic figure-ground segmentation, whenever a "moving" event was recorded in the video test library by the object's embedded sensors. Each detection algorithm was then constrained to detect only in areas with movement using the segmentation mask.

6.2 Results

The results of using context from movement sensing in multi-cue detection are shown in Figure 7 (right). Similar to the results in Figure 7 (left), the overall detection performance improves for all objects in the study with the use of multiple cues; however, this improvement is much greater with context information. Over all objects the mean average detection performance now increases to 88.72% of video frames, using all 4 cues. This is an improvement from the previous experiment of 50.83% and 45.26% over using just the single best cue and all 4 cues respectively. The most frequent combination of two best performing cues was now Colour and Texture, ranked as the best two for 40% of all objects. However, increasing the number of cues considered beyond the best two again gives little additional benefit (2.10% for 3 cues) and the addition of a fourth cue did not improve detection performance in any object.

6.3 Discussion

The results indicate that the use of context information from embedded movement sensing significantly increases detection performance for all cues and hence overall detection robustness for mobile objects. We infer that this is the result of the context information being used to mask the distracting and cluttered background. However, the limitation to using movement sensing alone is that when detecting static objects we can only mask moving areas in the image, hence the benefit will vary depending on the size of the area. Large moving areas provide similar performance to a moving object, whereas completely static scenes are equivalent to the multi-cue experiments.

7 Conclusion

In this paper we conducted three experimental studies to firstly understand impact of object scale and rotation on different detection methods, secondly to explore how the different detection cues combined for increased detection performance in real-world environments, and thirdly, to show the potential of combining detection with context information from embedded movement sensing.

This work provides three main contributions. We found that scale and rotation has a large impact on detection performance and that with the exception of the colour method we need to train our algorithms with multiple object viewpoints to detect an object in any pose. Similarly, we learned that detection performance varies when the algorithm is trained at different distances. This knowledge is significant, as it allows us to determine the best orientation and training distance for extracting appearance knowledge about the object from the camera image for initial object appearance training, and for the Cooperative Augmentation system to determine when to extract additional knowledge at runtime.

The second part of our contribution arises from the multi-cue experiment. Here we showed that for all objects the use of multiple cues in detection provides a performance benefit over using a single cue. None of the cues had high detection performance for all objects, underlining the fact that different objects require different appearance representations and detection methods, and validating the multi-cue approach proposed for the Cooperative Augmentation concept in [2]. The third part of the contribution is the finding that the use of context in detection from movement sensing in the object significantly increases detection performance, indicating the combination of different sensor modalities is important. Movement sensing constrains the search space by reducing distractions from the cluttered real-world environment.

More research is required on which vision algorithms are best suited to detecting objects and to better understand the role of context in detection – specifically, what impact different context information and other embedded sensors have on detection.

Acknowledgements. This research is supported by the EPSRC, the Ministry of Economic Affairs of the Netherlands through the BSIK project Smart Surroundings under contract no. 03060 and by Lancaster University through the e-Campus grant.

References

1. Strohbach, M., Gellersen, H.-W., Kortuem, G., Kray, C.: Cooperative Artefacts: Assessing Real World Situations with Embedded Technology. In: Davies, N., Mynatt, E.D., Siio, I. (eds.) UbiComp 2004. LNCS, vol. 3205, pp. 250–267. Springer, Heidelberg (2004)
2. Molyneaux, D., Gellersen, H., Kortuem, G., Schiele, B.: Cooperative Augmentation of Smart Objects with Projector-Camera Systems. In: Krumm, J., Abowd, G.D., Seneviratne, A., Strang, T. (eds.) UbiComp 2007. LNCS, vol. 4717, pp. 501–518. Springer, Heidelberg (2007)
3. Ehnes, J., Hirota, K., Hirose, M.: Projected Augmentation - Augmented Reality using Rotatable Video Projectors. In: Proceedings of the Third IEEE and ACM International Symposium on Mixed and Augmented Reality (ISMAR 2004), Arlington, VA, USA, September-October (2004)

4. Bandyopadhyay, D., Raskar, R., Fuchs, H.: Dynamic Shader Lamps: Painting on Movable Objects. In: Proc. of the IEEE and ACM International Symposium on Augmented Reality (ISAR 2001), New York (2001)

5. Borkowski, S., Riff, O., Crowley, J.L.: Projecting rectified images in an augmented environment. In: IEEE International Workshop on Projector-Camera Systems (PROCAMS 2003), Nice, France, October 12 (2003)

6. Pinhanez, C.S.: The Everywhere Displays Projector: A Device to Create Ubiquitous Graphical Interfaces. In: Abowd, G.D., Brumitt, B., Shafer, S. (eds.) UbiComp 2001. LNCS, vol. 2201. Springer, Heidelberg (2001)

7. Swain, M.J., Ballard, D.H.: Color indexing. International Journal of Computer Vision 7(1), 11–32 (1991)

8. Jurie, F., Dhome, M.: Hyperplane Approximation for Template Matching. IEEE Transactions on Pattern Analysis and Machine Intelligence 24(7), 996–1000 (2002)

9. Schiele, B., Crowley, J.L.: Recognition without Correspondence using Multidimensional Receptive Field Histograms. International Journal of Computer Vision (IJCV) 36(1), 31–50 (2000)

10. Murase, H., Nayar, S.K.: Visual Learning and Recognition of 3D Objects from Appearance. International Journal on Computer Vision 14(1), 5–24 (1995)

11. Turk, M., Pentland, A.: Eigenfaces for Recognition. Journal of Cognitive Neuroscience 3(1), 71–86 (1991)

12. Belongie, S., Malik, J., Puzicha, J.: Shape Matching and Object Recognition Using Shape Contexts. IEEE Transactions on Pattern Analysis and Machine Intelligence 24(4), 509–522 (2002)

13. Mokhtarian, F.: Silhoutte based Isolated Object Recognition through Curvature Scale. IEEE Transactions on Pattern Analysis and Machine Intelligence 17(5), 539–544 (1995)

14. Lowe, D.G.: Distinctive Image Features from Scale-Invariant Keypoints. International Journal of Computer Vision 60(2), 91–110 (2004)

15. Mikolajczyk, K., Schmid, C.: A Performance Evaluation of Local Descriptors. IEEE Transactions on Pattern Analysis and Machine Intelligence 27(10), 1615–1630 (2005)

16. Mikolajczyk, K., Tuytelaars, T., Schmid, C., Zisserman, A., Matas, J., Schaffalitzky, F., Kadir, T., Gool, L.V.: A Comparison of Affine Region Detectors. International Journal on Computer Vision 65(1-2), 43–72 (2005)

17. Spengler, M., Schiele, B.: Towards Robust Multi-cue Integration for Visual Tracking. In: Proceedings of the Second International Workshop on Computer Vision Systems. Springer, Heidelberg (2001)

18. Li, P., Chaumette, F.: Image Cues Fusion for Object Tracking Based on Particle Filter. In: Perales, F.J., Draper, B.A. (eds.) AMDO 2004. LNCS, vol. 3179, pp. 99–107. Springer, Heidelberg (2004)

19. Brasnett, P., Mihaylova, L., Canagarajah, N., Bull, D.: Particle Filtering with Multiple Cues for Object Tracking in Video Sequences. In: SPIE's 17th Annual Symposium on Electronic Imaging, Science and Technology, San Jose California, USA, pp. 430–441 (2005)

20. Giebel, J., Gavrila, D.M., Schnörr, C.: A Bayesian Framework for Multi-cue 3D Object Tracking. In: Pajdla, T., Matas, J(G.) (eds.) ECCV 2004. LNCS, vol. 3024, pp. 241–252. Springer, Heidelberg (2004)

21. Lindeberg, T.: Scale-Space for Discrete Signals. IEEE Transactions on Pattern Analysis and Machine Intelligence 12(3), 234–254 (1990)

22. Harris, C., Stephens, M.: A combined corner and edge detector. In: Proceedings of the 4th Alvey Vision Conference, pp. 147–151 (1988)

Design and Evaluation of a Sound Based Water Flow Measurement System

Alejandro Ibarz[1], Gerald Bauer[2], Roberto Casas[1], Alvaro Marco[1], and Paul Lukowicz[2]

[1] TecnoDiscap Group, University of Zaragoza,
Maria de Luna 3, 50018 Zaragoza, Spain
[2] Embedded Systems Lab, University of Passau,
Innstr. 43, 94032 Passau, Germany
aibarz@unizar.es, gerald.bauer@uni-passau.de, rcasas@unizar.es,
amarco@unizar.es, paul.lukowicz@uni-passau.de
http://www.wearable-computing.org

Abstract. This paper presents a low-cost, easy to install sound-based system for water usage monitoring in a household environment. It extends the state of the art but not only detecting that water is flowing in a pipe, but also quantifying the flow thus allowing us to compute the amount of water used. We describe the system architecture including hardware, software and the signal processing and pattern recognition algorithms used. We present an extensive evaluation in a real life noisy kitchen environment. We show an accuracy of over 90 percent on classifying six different water flow levels. We also demonstrate good performance measuring water consumption when compared with the home's water meter.

Keywords: Smart-sensors models, sensing at home, acoustic event classification, water usage monitoring.

1 Introduction

The recognition of every day household activities is an important component of many pervasive computing applications. This includes so called ambient assisted living systems that aim to help handicapped and elderly people lead an independent life at home rather than be admitted to a nursing home. The work described in this paper is part of a large European Union funded project directed at such assistive applications (MonAmi, www.monami.info). Within the project our groups work on every day activity tracking systems that are based on a combination of coarse location [1], auditory scene analysis [2] and appliance use. In this paper we describe a recent extension of our system that allows us to recognize the amount of water flowing through a particular pipe using a cheap, low quality off the shelve microphone module ('scavenged' from a Bluetooth Headset).

The use of microphones to detect water flow and the relevance of the information in detecting a range of every day activities has been demonstrated by Fogarty et al. [3]. We

D. Roggen et al. (Eds.): EuroSSC 2008, LNCS 5279, pp. 41–54, 2008.

extend this work by detecting not just the fact that water flows through a particular pipe, but also quantifying the flow speed. This allows us to calculate the total amount of water used for a particular activity. The work is motivated by the observation that the amount of water used (as well as the flow speed) is a potentially useful feature to distinguish activities with otherwise similar signature. Thus getting a glass of water to drink involves a different amount of water then filling a large pot for cooking noodles. We also expect the ability to detect flow speed to allow us to distinguish different water usage sources with only a single microphone located at a common pipe rather than having to place a microphone at every individual tap (as done by Fogarty et al.[3]).

Clearly the question how the water flow information can be best used in activity recognition requires further research and we aim to undertake this research at a later stage. However, before doing so, we wanted to investigate whether such detection is possible with low quality cheap microphones. The paper describes the results of this investigation. We present the architecture of our system, the machine learning techniques used to recognize the water levels and an extensive empirical evaluation. For evaluation the system is deployed in a kitchen of a shared flat, in which four people are living in. Different experiments are performed to examine the feasibility of the system. The kitchen ambient noise is taken into account when designing the system, due to the fact that pipes are excellent conductors of sound. A model of the kitchen's water tap flow is obtained via machine learning and is used to evaluate the system's performance. Furthermore, for every experiment the system's calculation of the used water amount is compared with the measurement of the home's water meter.

1.1 Related Work

Among all the work done in Acoustic Event Detection/Classification, we focus on applications that detects water usage in the home environment.

Chen et al. [4] present an automatic acoustic bathroom monitoring system. The system is designed to recognize and classify different activities of daily living occurring in a bathroom based on sound. All the events detected are related to water usage. It uses a Hidden Markov Model classifier and Mel Frequency Cepstral Coefficients features. Preliminary results showed high average accuracy. The system consists on a microphone wired to a processing unit. In contrast, our approach uses wireless sensors and the possibility to determine water flow measurement monitoring.

Kraft et al. [5] describes feature extraction methods for sound classification. They record high quality audio in a kitchen environment and detect and classify kitchen acoustic events including water.

Stäger et al. [6] presents a prototype of a wearable sound-analysis based, user activity recognition system. Even though they focus on low-power and a tradeoff analysis between recognition performance and computation complexity, they evaluated the performance of the system in a kitchen environment. This reported good recognition results including water tap usage.

Fogarty et al. [3] describe an excellent work about a low-cost, unobtrusive and easy to install approach for water usage monitoring based on sound in a home environment. They deploy low-cost wireless microphone-based sensors attached to the outside of existing pipes at critical locations. The system is able to recognize many activities related to water usage like cloth washer, dishwasher and shower usage as well as toilet and bathroom sink activities longer than ten seconds.

1.2 Paper Contribution

The key contribution of the paper is to show that low quality, low cost microphones attached to water pipes can detect not only the fact that water is flowing, but also quantify the amount of water and to demonstrate this in an extensive empirical evaluation in a noisy, real life environment. We begin by describing a detailed concept of the system. Next we describe the system's architecture followed by the detailed system's implementation. Then we discuss the system's evaluation in a kitchen located in a shared flat environment (four people are living there), describing the whole process of modeling the installation and validating the results.

2 System Overview

As sensor we use a commercial, off the shelve Bluetooth audio device. The device is placed in a small box covered with foam (see Fig. 1). Then the box is attached to the inflow pipe of a water tap (see Fig. 2). So a big advantage is that neither pipes nor the water tap itself has to be removed. Furthermore, it only requires a screwdriver to close the small box. As a result the sensor installation is not time-consuming and no technical specialist is required which saves costs. Thus, the system is easy to install and unobtrusive. Although our system is battery operated, many places where we could access water pipes also have a nearby socket (e.g. under the kitchen sink). The audio device sends the sound information to a computing unit where audio data processing is performed. Because our system uses a wireless sensor there is no need for the user to install a cable to the processing unit which means less intrusion into the user's environment. Fig. 3 shows the architecture of our system. Every time new audio data arrives, the processing unit will process them using a chain of algorithms, which are implemented in the CRN Toolbox [7].

Fig. 1. Headset microphone sensor placed in an isolated box

Fig. 2. Sensor mounted at an inflow pipe of a washbowl

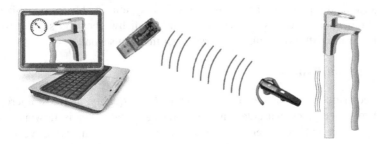

Fig. 3. System architecture description

Table 1. Detailed information of the different running water levels

Water level	Flow value (ml/s)
1	7.66
2	34
3	56
4	90
5	125
6	145

Audio data were previously analyzed to design the algorithms for getting useful information for further machine learning. Once the right features are obtained, a model of the installation is created. The installation's model is created by an initialization step, where training data of silence and also of six different levels of water flow (see table 1) without noise were recorded. To be also resistant against environment noise different rules have been defined and the system was evaluated using rules and also without rules.

The results of the system are information about the status of a water tap – water running or not – and in case of running water the closest learned water level. This information might be useful for activity monitoring and also for water consumption measurement.

3 System Implementation

3.1 Sensor

The sensor used for audio acquisition is a commercial off the shelve Bluetooth audio device (Headset). The case and the speakers were removed and the microphone was placed in a small box covered with foam (see Fig. 1 and 2). Two holes were made in the box to fit the flexible pipe and the box cap is closed with screws to confer the sensor some noise isolation to the environmental noise.

3.2 Algorithms and Software

First we give a short introduction about the CRN Toolbox to clarify why we chose this software tool for our system implementation. Afterwards we show a detailed description of the used data processing steps and finally we compare different classifier models towards their ability to fulfill our objective to recognize running water and to measure the about amount of used water. Experiments performed in this section were carried out using the Rapid Miner tool [8].

3.2.1 CRN Toolbox

We choose the Context Recognition Network (CRN) Toolbox [7] (available under LGPL from http://crnt.sf.net) to implement the above described system. The CRN Toolbox is a software tool optimized for the implementation of multi-modal, distributed activity and context recognition systems running on POSIX operating systems. It contains a broad set of fundamental ready to use algorithms (called tasks) for different common applications like data classification and filtering as well as special applications like sound and image processing. Because all tasks are stand-alone and provide a generic interface, complex data processing chains can be created with ease by interlinking needed tasks. The main reason for choosing the CRN Toolbox for our application was beside the chance of using currently efficient implemented tasks for sound data processing and the possibility to implement new tasks with ease, to have a platform independent application. So, different devices like PDAs and mainstream computer systems can be used to run our application without any modifications.

3.2.2 Data Processing Chain

We use a chain of Toolbox tasks to recognize different levels of water, to calculate the amount of used water and to separate them also from other sounds like speech and device noises. An overview about all used tasks is shown in figure 4. There the Bluetooth Microphone reader, the Fast Fourier Transformation, some classifier and also the ResultWriter were already implemented in the Toolbox. Step one to step four are very common in sound recognition applications and will be introduced very shortly in the next section. In contrast to that we will focus more detailed on the used classifiers and the additional filter tasks in section 3.2.3 and 3.2.4. In the following, a detailed description of each task is given.

Bluetooth Microphone Reader. This task connects to the Bluetooth microphone sensor and reads its raw audio information using synchronous connection-oriented (SCO) Bluetooth logical transport [9]. The master can create up to three SCO links to different slaves. Even though at this moment the system uses only one audio device, it could get audio data from 3 different audio devices improving the scalability of the system.

Fast Fourier Transformation (FFT), Mel Frequency, Log10. Using FFT, Mel Frequency and Log10 the system performs the feature extraction for the classification. First, digital values delivered from the Bluetooth microphone are processed using Fast Fourier Transformation (FFT) to get a spectrum of the current sound signal. Because our objective is to recognize continuous sound of running water and the spectral signatures of the different levels were slightly varying in some cases, we chose 8192-point FFT for improved frequency resolution and rectangular windowing with no overlapping. As result we obtain one sound data spectrum per second. In order to reduce the spectrum's size we map the FFT's frequency values to the "mel scale" using 31 triangular overlapping windows following a logarithmic pattern [10], a high number of features were obtained to keep good frequency resolution. As result we obtain a vector of 31 features whereas on each feature a logarithm function to base 10 is performed.

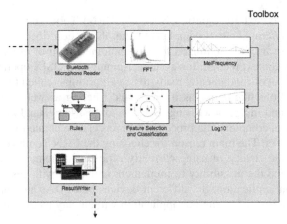

Fig. 4. Chain of used Toolbox tasks

Feature Selection. To avoid overfitting and to reduce computing cost we perform a dimension reduction. Therefore training data is analyzed using Information Gain (IG) and Gini Index (GI) as feature weighting methods at an initial step [11][12]. Only the seven most important features will be considered in the following processing steps.

Classification. The above selected features are used as input for different classifier. Therefore we choose the following classification models: Decision Tree [13], k nearest neighbor (kNN) [13] and Support Vector Machine (SVM) [14]. A detailed description of evaluation results can be found in section 3.2.3.

Rules. To reduce false classifications and to detect surrounding noise sounds (classifier models are not trained to recognize surrounding noise) we have defined a set of rules, which is applied to the result of the former classification task (see section 3.2.4).

ResultWriter. The result writer is used to make the systems output available to other devices.

3.2.3 Feature Ranking and Data Classification

As mentioned above we chose a subset of features to increase generalization ability and to reduce computing cost. Figure 5 shows Mel-filter values of different water flow level and also silence. It can be seen that Mel-filter channels from 6 to 20 might

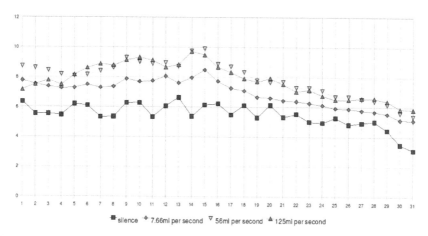

■ silence ◆ 7.66ml per second ▽ 56ml per second ▲ 125ml per second

Fig. 5. Examples for different constant water flows measured in ml per second (x-axis = Mel-filter channels, y-axis = Mel-filter value). Experiments showed that even for constant flows Mel-filter channels lower than 5 and higher than 22 vary very strongly.

Table 2. Detailed frequency information (Hz) of important feature channels

Channel	Start	Center	Stop
7	308	370	437
8	370	437	508
9	437	508	583
10	508	583	663
11	583	663	747
14	837	932	1033
20	1504	1640	1784

Table 3. Classification performance results and settings for different models

Classifier	Settings	Result (10-fold CV)	Result (Deviant data)
Decision Tree[1]	max depth = 10	99.29% +/-0.96	66.11%
	max depth = 5	97.29% +/-2.07	65.45%
kNN[2]	K=10	100%	90.70%
	K=5	100%	90.37%
SVM	RBF Kernel, C = 0, gamma = 1/7	97.57% +/-0.91	82.72%

[1] The Information Gain was used as criterion for numerical splits.
[2] The Euclidian distance was used as distance measure.

be dedicated for distinguishing between the different water levels. Comparing the 10 most weighted features for both the IG and GI and choosing the most important seven ones our first assumption is confirmed. Channels 7, 8, 9, 10, 11, 14 and 20 are ranked as very important features (see Table 2). Therefore we will consider only these channels for further processing tasks. Because our aim was to distinguish between silence

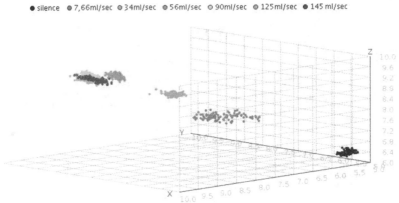

Fig. 6. Visualization of learning data using channel 7 (x-axis), 8 (y-axis) and 9 (z-axis)

and six different levels of running water (see Table 1)[1], we have recorded 100 training samples for each class. To ensure generalization ability we used a 10-fold cross-validation with stratified sampling [13] and we recorded 50 additional samples for each water level to evaluate the trained system. Because our intention was also to check how classifier handles sound data, which belong to water levels nearby recorded constant flows, we deviate a little from the trained water flow levels when recording the 50 additional data. Table 3 shows the results for each classifier and its settings.

It can be seen that using a 10-fold cross validation each model is able to classify different levels of water very well. In contrast to that using deviant data for model evaluation the classification performance for each model decreases very strong. Only the kNN is able to classify sufficiently with an accuracy of about 90%. We have analyzed the classification results for each class to find the weak points. All classifiers are able to distinguish between silence and water flow levels 7.66ml/sec, 34ml/sec and 145 ml/sec very well (about 90%). However all fails in distinguishing between water levels 56ml/sec, 90ml/sec and 125ml/sec. Figure 6 visualizes sound data using three of seven features and show that these classes are placed very close to each other, which might be the reason for that. But of course using all seven features the location of shown sound data could be very different.

Because the Decision Tree and also the SVM adapt too strong to training data and are not able to allot near located data to belonging classes, we consider only the kNN as classifier for our application.

3.2.4 Rules
Up to now we have neglected noise sounds like speaking person and sounds from devices, which are located in the washbowl's environment. Because the kNN is trained using only silence and water sound data, every time a noise sound appears it

---

[1] To measure the amount of water for a constant water flow we used the flat's main water counter.

will be assigned to one of these classes. The easiest way to classify environment sound as noise would be to train the kNN also with noise sound data. But in real applications it is neither comfortable nor possible for a user to record all possible noise sounds once after the sensor installation. So we defined a distance measure, which is used to decide whether the kNN classification for running water is accepted or in case of surrounding noise neglected (noise reduction). The distance measure is defined on the equation 1.

$$isWaterSound = \begin{cases} 0, & \text{if } (kNN\_dist > 40) \\ 1, & \text{otherwise} \end{cases} \qquad (1)$$

There *kNN_dist* is the sum of distances between the current complete feature vector and all of the *k* nearest neighbors which belongs to the winning class. If the kNN classifies a sound sample of type running water (any type of water level) the classification result will only be accepted if *isWaterSound* is equal 1 and the improved result is the recognized class of water. Otherwise the sound sample is labeled as surrounding noise. Thereby the threshold for *kNN_dist* was chosen by analyzing experimental results.

But even if the system is able to recognize noise sounds, the fact that very loud surrounding noise drowns the running water signal is still a problem for calculating the used amount of water. Therefore we defined a set of rules, which approximate the used water consumption during overlapping surrounding noise and reduce also false classifications (some noise sounds are very similar to water sounds and cannot rejected by the distance measure). The rules are defined as following:

- To reduce the impact of false classifications running water is only recognized if during the last three seconds sound data was classified only as running water. So single outliers are neglected.
- To avoid single false classifications by distinguishing between different levels of water, the last five recognized ones are considered and the most present water level is assumed to be the current water level.
- If running water was recognized and it will be interrupted not longer than ten seconds by a continuous noise, the system assumed the last water level as current water level. If there are more than ten seconds of overlapping noise sounds the system will assume that there is no longer running water.
- If running water is interrupted by silence the system assumes that the water tap has been closed and there is no longer running water.

Of course some assumptions may lead to false results, e.g. if there is overlapping environment noise for several minutes and the water tap will be used during that time the system is not able to recognize the running water. Therefore the next section includes several real time experiments to evaluate the systems quality using rules and also without the introduced rules.

4 System Evaluation

To evaluate the quality of the system we performed different types of experiments. The results and a detailed experiment description are shown in the next sections.

There all experiments have been performed in a kitchen environment (see Fig. 7) and the sensor was mounted on the cold-water inflow pipe of a washbowl.

4.1 Water Flow Measurement

The first experiment is used to validate how accurate the introduced system is able to measure the used amount of water. Therefore we have performed three different experiments. First we calculated the amount of used water during a period in which no surrounding noise occurs. Second we measured the amount of used water while surrounding noise was existent and last we performed only environment noise to see if the system is able to neglect these sounds. As noise we considered all typical sounds, which can occur in a kitchen environment, like speaking and laughing, opening/closing doors and shelves, using devices like mixer, microwave and fume hood, listening to music and putting/removing dishes to/from the sink. To have a reliable comparison we used the main water flow meter of the flat as ground truth. The results are shown on Fig. 8 whereas the kNN, the kNN with noise reduction and also the whole system (kNN, noise reduction, and rules for reducing false classifications) were used to distinguish between the different levels of water. In doing so we sum-

Fig. 7. Kitchen environment

marized for each approach the classification results to get an approximation of the whole amount of used water.

Results for the kNN without any further processing steps are quite good for water measurement applications without noise. Because loud noise is recognized as water the calculated amount of water is much higher than the used one for the second experiment. Using the kNN and an additional noise reduction it can be seen that the amount of calculated water is lower than the real one for experiment one and two. This means that some sounds of running water have been classified as noise and so values for running water got lost. In contrast to that using the kNN with distance measurement and applying rules the calculated water amount for these two experiments improves the result of the kNN with noise reduction. The results for both experiments are quite well but it was not possible to measure the exact amount of water. The reasons for this are that on the one hand the system tries to estimate the water level during overlapping noise (rule three) and on the other hand the system needs some seconds to recognize the new water level after a change (rules one and two). Therefore information about the current water level gets lost. Of course we have to take also into account that we use only six different levels of water to approximate the amount of used water.

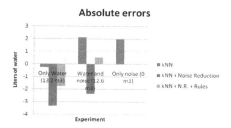

Fig. 8. Experiment results for different methods and environments. Real flow of water was 13.2 m³ in "Only water" setup and 12.6 m³ in "Water and noise" setup. "Only noise" setup has not water flow.

Having a look at the last experiment it can be seen that the kNN fails. Because no noise training data is available the kNN has to allot noise sounds to silence or water. So nearly two liter of used water has been calculated for activities without water usage, which makes this model unusable in practical applications. The kNN with additional distance measurement is quite accurate. Only 0.02 liters of used water have been recognized. In contrast to both the whole system is able to classify all noise sounds correctly and so no running water was recognized. Taking all three experiments into account it can be seen that the whole system works quite well in each case.

Furthermore, the system was evaluated without noise using only hot water (maximum flow) and a mixture of hot and cold water (50% of each one and maximum flow) while keeping the device on the cold water pipe. This was performed to test how the use of hot water could modify the classification and therefore the amount of cold water measured. Results shown on Fig. 9 show a good system performance when only hot water or a mixture is used.

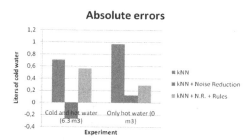

Fig. 9. Experiment results for different methods and environments. Real flow of water was 6.3 m³ in "Cold and hot water" setup. "Only hot water" setup has not water flow.

4.2 Common Water Flow measurement: Preparing Food

The second experiment focuses on a real application. So every type of water level is used and our objective is to see how well the system is able to approximate the used amount of water knowing only six different water levels. Therefore one test person, who was not informed about how the system works exactly, was asked to prepare pasta with ham and tomato sauce, toast with salmon pieces and to clean the dishes after that. All in all it took about 35 minutes to finish the given task. During this time a lot of environment noise was produced. The user opened and closed many times the fridge and drawers, he used different utilities for preparing the food like knifes for cutting ham and salmon, and he listened to music and made phone calls. After the food was finished he put all used dishes into the sink and cleans them. As before our system calculates the used amount of water during this period. Figure 10 shows the experiment result.

At all 18.2 liters of water have been used. Having a look at the kNN results it can be seen that the calculated amount of water is higher. Because - as mentioned before - the kNN will also classify noise as water this kind of deviation was expected. As explained in 4.1 the kNN with noise reduction calculates again lower values than the whole system and also than the real amount of used water. In contrast to 4.1 the whole system was not able to reach the same quality again. The reason for this is that in applications like preparing food the water tap is used very often for a short time lower than three seconds (e.g. for washing hands or cleaning a knife).

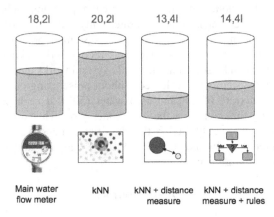

Fig. 10. Water flow measurement during a kitchen application (preparing food and washing dishes)

Because of rule one such short water tap usages are neglected. Reducing the threshold for the first rule will solve this problem but it should be concerned that a lower threshold will also benefit the amount of false classifications. So it is up to the user to choose an appropriate threshold for his specific application. At all it can be summarized that the amount of used water - using only six different levels of water - can be approximated quite well using the whole system.

4.3 Common Observations and Universal Usage

During all experiments we could see that the system had problems to recognize running water lower than 7.66 ml per second. Most of the time, such a low water level was classified as silence. This problem might be solved if the system is trained with sound data of this water level too.

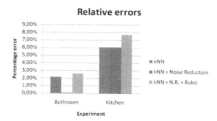

Fig. 11. Experiment results for different methods and environments

We have also tried to use the trained system in other environments like bathroom and public toilets. Because washbowls may use different material for inflow pipes the water sounds may be different, therefore the system must be trained for every installation. But using the introduced system a self-acting sensor training to a specific location can be performed very easy. Because of using the kNN classifier only sounds for different levels of water and also silence must be recorded once. Experiments showed that there is no need to change selected features or thresholds, which are used by both the noise filter and the rule layer. Furthermore, the system performance showed good results when evaluated on two different installations without environmental noise as can be seen on Figure 11.

5   Conclusion and Future Work

This paper proposes a low-cost, easy to install, unobtrusive, wireless sensor-based system to detect water usage activity patterns in a kitchen environment. It is able to distinguish between six different levels of running water through a pipe with very good accuracy. Furthermore it is shown that is was also possible to approximate the amount of used water and shows a good performance despite considerable levels of noise. The system needs to be trained once after installation by recording only sound data from six different levels of water and silence.

In future work we investigate how to use best the water flow speed and amount information in complex activity recognition. To this end we will combine the system described in this paper with our indoor sub room level location system [1], socket based power usage trackers produced by one of the MonAmi partners and motion information form a body worn accelerometer. We will also investigate how auditory scene analysis algorithms demonstrated in previous work [2] can be added to the system so that other kitchen noises are not discarded as noise, by used as relevant activity information.

Acknowledgments. This work was supported by the European Union, under projects MonAMI and EasyLine+.

References

1. Bauer, G., Lukowicz, P.: Developing a Sub Room Level Indoor Location System for Wide Scale Deployment in Assisted Living Systems. In: Miesenberger, K., Klaus, J., Zagler, W.L., Karshmer, A.I. (eds.) ICCHP 2008. LNCS, vol. 5105, pp. 1057–1064. Springer, Heidelberg (2008)

2. Stäger, M., Lukowicz, P., Tröster, G.: Power and accuracy trade-offs in sound-based context recognition systems. Pervasive and Mobile Computing, vol. 3, pp. 300–327. Elsevier, Amsterdam (2007)
3. Fogarty, J., Au, C., Hudson, S.E.: Sensing from the Basement: A Feasibility Study of Unobtrusive and Low-Cost Home Activity Recognition. In: Proceedings of the ACM Symposium on User Interface Software and Technology (UIST 2006), pp. 91–100 (2006)
4. Chen, J., Zhang, J., Kam, A.H., Shue, L.: An automatic acoustic bathroom monitoring system. In: IEEE International Symposium on Circuits and Systems, 2005. ISCAS 2005, May 23-26, 2005, vol. 2, pp. 1750–1753 (2005)
5. Kraft, F., Schaaf, T., Waibel, A., Malkin, R.: Temporal ICA for Classification of Acoustic Events in a Kitchen Environment. In: 9th European Conference on Speech Communication and Technology 2005, Interspeech 2005, Lisboa, Portugal, September 13, 2005, pp. 2689–2692 (2005)
6. Stager, M., Lukowicz, P., Troster, G.: Implementation and evaluation of a low-power sound-based user activity recognition system. In: Eighth International Symposium on Wearable Computers. ISWC 2004, 31 October-3 November 2004, vol. 1, pp. 138–141 (2004)
7. Bannach, D., Amft, O., Lukowicz, P.: Rapid Prototyping of Activity Recognition Applications. Pervasive Computing 7(2), 22–31 (2008)
8. Mierswa, I., Wurst, M., Klinkenberg, R., Scholz, M., Euler, T.: YALE: Rapid Prototyping for Complex Data Mining Tasks. In: Proceedings of the 12th ACM SIGKDD International Conference on Knowledge Discovery and Data Mining (KDD 2006), pp. 935–940 (2006)
9. Bluetooth Special Interest Group, https://www.bluetooth.org/apps/content/
10. Stevens, S., Volkman, J., Newman, E.: A scale for the measurement of the psychological magnitude of pitch. Journal of the Acoustical Society of America 8(3), 185–190 (1937)
11. Liu, H., Motoda, H.: Feature Extraction, Construction and Selection. Kluwer Academic Publisher Group, Boston (1998)
12. Liu, H., Motoda, H.: Feature Selection for Knowledge Discovery and Data Mining. Kluwer Academic Publisher Group, Boston (1998)
13. Bishop, C.: Pattern Recognition and Machine Learning, 1st edn. Springer, Heidelberg (2007)
14. Cristianini, N., Shawe-Taylor, J.: An Introduction to Support Vector Machines and Other Kernel-Based Learning Methods. Cambridge University Press, New York (2000)

Gaussian Process Person Identifier Based on Simple Floor Sensors

Jaakko Suutala[1], Kaori Fujinami[2], and Juha Röning[1]

[1] Intelligent Systems Group, Infotech Oulu
P.O. Box 90014 University of Oulu, Finland
{jaska,jjr}@ee.oulu.fi
[2] Department of Computer, Information and Communication Sciences
Tokyo University of Agriculture and Technology, Japan
fujinami@cc.tuat.ac.jp

Abstract. This paper describes methods and sensor technology used to identify persons from their walking characteristics. We use an array of simple binary
switch floor sensors to detect footsteps. Feature analysis and recognition are performed with a fully discriminative Bayesian approach using a Gaussian Process
(GP) classifier. We show the usefulness of our probabilistic approach on a large
data set consisting of walking sequences of nine different subjects. In addition,
we extract novel features and analyse practical issues such as the use of different
shoes and walking speeds, which are usually missed in this kind of experiment.
Using simple binary sensors and the large nine-person data set, we were able to
achieve promising identification results: a 64% total recognition rate for single
footstep profiles and an 84% total success rate using longer walking sequences
(including 5 - 7-footstep profiles). Finally, we present a context-aware prototype
application. It uses person identification and footstep location information to provide reminders to a user.

Keywords: Person identification, machine learning, floor sensors, context-
awareness.

1 Introduction

Providing context-aware services to the user by means of smooth human-computer interaction requires natural and transparent ways to identify and locate users [1], [2], [3].

We present an approach to person identification based on human motion. More
specifically, we concentrate on a person's style of walking, which presents behavioral
characteristics of biometrics and is very natural because no additional action is required
of the user. In this work, binary switch sensors are used to detect a person s walking
sequence. A binary switch sensory system consisting of an array of 300 sensors was
installed on the surface of the floor. Each 10 cm x 10 cm binary switch senses weight
affecting its surface. A 3 m^2 floor area was covered to collect data and to recognize
walkers.

User identification is based on statistical machine learning. We use a Gaussian process (GP) classifier [4] with specific features extracted from the footstep profiles produced by the sensor array as well as features calculated between consecutive footsteps.

D. Roggen et al. (Eds.): EuroSSC 2008, LNCS 5279, pp. 55–68, 2008.

The usefulness of the GP as a fully Bayesian kernel method relies on the ability to model uncertainty of data, which leads to automatic determination of hyperparameters (e.g., the importance of different features). It also produces conditional posterior probabilities of class labels (i.e., the degree of belonging to a certain class), which allows extensions and different post-processing capabilities of the classifier. Feature extraction itself is based on the standard methodology of image processing, as the sensor array can be presented as a binary image. Different features are extracted from the binary image based on single footstep profiles (e.g., length and width calculated from connected components in the binary image) and from the sequence of walking (e.g., step length and duration between consecutive footstep profiles). Along with these typical walker identification features we examined more specific ones that were calculated from the time-integrated signal. In practice, the binary signals were summed over time to form a grey-level image, and then features such as mean, standard deviation, and the center of mass were extracted from the connected components.

Data sets were collected from nine different subjects, including 20 walking sequences for each person. Each person wore their own shoes. In addition, four persons walked at three different walking speeds (slow, normal, fast) and also with two different pairs of shoes and without shoes. In the GP classification we examined the identification based on single footstep profiles as well as the identification using information from walking sequences (i.e., including multiple footstep profiles). The importance of different features were also analyzed.

This paper presents a simple yet modular floor sensor system that is able to identify persons based on their walking. It also describes an accurate classification method that is simultaneously able to analyze and choose the most important features and produce posterior probability of ID labels for post-processing. Furthermore, the effect of changes in walking speed and footwear is analyzed for the first time.

2 Related Work

Various floor sensor settings have been used to model human behavior for identification, tracking, and other purposes in smart and interactive environments as well as in health care. In the early works by [5] and [6], footstep identification was based on a small area of ground reaction force (GRF) sensors using nearest-neighbor and hidden Markov model (HMM) methods, respectively. In the work by [7], human GRF-based authentication system was developed for use as part of a surveillance system. Recently, a sensor installation, collection of a large data set and experiments with a person verification scenario were presented in [8]. They used a GRF sensor with geometric and holistic features along with a support vector machines classifier.

In [9], electromechanical films (Emfi) that measures dynamic pressure changes on the floor surface were used for person identification. A comparison of different methods (e.g., support vector machines and neural networks) was done. In addition, classifier fusion techniques were applied to combine different feature sets and walking sequence information to achieve a more reliable recognition system.

UbiFloor [10], uses simple ON/OFF switch sensors, and identification is based on features of both single footsteps and walking calculated from five consecutive footsteps on the floor. The sensor arrangement differs from our work, but the use of simple binary sensors is most similar to ours from the application viewpoint. A multi-layer perceptron (MLP) neural network was used as a classifier. [11] developed a high-resolution low-cost pressure sensor mat made of resistive switches. They also performed person identification based on sequential features such as stride length, gait period, and heel-to-toe ratio along with an Euclidean distance measure as a classifier.

A sensor approach similar to this work was established by [12]. However, they concentrated on human tracking applications based on Markov chain Monte Carlo methods. [13] presents a system that also uses binary ON/OFF sensors in which over 65,000 pressure switches in an area of 4 m^2 give a very high resolution to the modeling of the details of single footstep profiles as an image of footprints. The floor was tested by detecting humans and robots and discriminating between them. [14] reported the use of a beneath-the-floor accelerometer and tactile sensors to model footsteps and footprints in order to recognize gender. [15] covered the floor of an interactive space with hexagonal pressure-sensitive floor tiles to detect the presence of users.

Besides identification and tracking, force plates have been used to detect and classify simple human body movements, such as crouches and jumps as well as standing up and sitting down [16]. A lot of work has also been done in medical research domains, including [17], where pattern recognition methods were used to classify different gait-related injuries based on GRF sensor measurements.

In summary, this work present a unique sensor approach to person identification ([12] uses similar sensors but they are used in a tracking application). We also extract novel features from the floor and analyze the importance of individual features as well as the effect of walking speed variations and different footwear, which are typically not included in the other studies related to floor-based identification. Our approach has a direct possibility of combining sequential information from multiple footsteps based on the classifier s posterior probability outputs. This is quite similar to [9], except that further post-processing is not needed to get the confidence of labels.

3   Binary Switch Floor Sensor System

VS-SF55 InfoFloor sensor system made by Vstone Corporation (in Japan) [18] was installed in our research laboratory. The system contains 12 blocks of 50 cm x 50 cm sensor tiles. Each tile includes 25 10 cm x 10 cm binary switch sensors. A 3 m^2 area was covered by altogether 300 sensors (see Fig. 1). The sensors use diode technology and are able to detect over 200-250 g/cm^2 weight affecting the surface. Data were collected from each sensor using a 16 Hz sampling rate and sent to a PC via an RS-232 serial interface. In the PC, a multi-threaded TCP-IP server was implemented to share raw sensor data with client applications.

Compared with other floor sensor technologies (e.g., Emfi [9]), the advantages of using this kind of floor sensor system are low cost, easy installation, and little need for pre-processing to get the data (e.g., for positioning and identification). Moreover, the sensor floor utilized in this paper is designed to be modular, which allows the sensor area to be able extended incrementally. On the other hand, compared with cameras, audio, or RFID technology, floor sensors are more stable, i.e, they do not suffer from environmental changes. A drawback is that only very limited information is obtained from the binary floor compared with cameras or other floor sensor technologies (e.g., Emfi and GRF sensors). This is very challenging, especially in complex recognition tasks such as person identification, where discrimination between different persons can depend on very detailed differences in persons walking styles. One aim of this work was to be able to extract such useful and discriminative information from this limited, yet practical, sensor system.

Fig. 1. Arrangement of binary sensor tiles

4   Discriminative Bayesian Classification: Gaussian Processes

Discriminative learning is a very effective way to train mappings from multidimensional input feature vectors to class labels. Kernel methods, in particular, have become state-of-the-art, due to their superior performance in many real-world learning tasks. Along with the popular support vector machines (SVM) [19], Gaussian processes (GP) [4] have recently been given much attention in the machine learning community.

Although the SVM method has many favorable properties, such as good generalization by finding the largest margin between classes, the ability to handle non-separable classes via soft-margin criteria, non-linearity modeling via explicit kernel mapping, sparseness by presenting data using only a small number of support vectors, and global convex optimization with given hyperparameters, it lacks some properties. One drawback of SVM is that it is directly applicable only in two-class problems. Thus, there have been various attempts to generalize it for multi-class classification. The simplest and most popular methods are based on multiple binary classifiers using one-vs.-one or one-vs.-rest approaches as well as error correcting output codes and directed acyclic graphs, to name a few [20].

Another problem is the choice of a good model, which is very important in kernel-based discriminative learning. This is due to the fact that a good solution is usually dependent on a number of hyperparameters (which control the properties of kernel mapping). In SVM, the hyperparameters (and possibly the good subset of features) need to be found using ad-hoc methods such as cross-validation or other search-based methods. When the number of hyperparameters or the number of features increases, the search space can become very large. Finally, SVM cannot directly give a confidence measurement as an output, it only gives a decision as an unscaled distance from the margin in the feature space. Posterior distribution over predicted class labels is a very important property in many pattern recognition systems in order to be able to implement some post-processing tasks (e.g., rejecting unreliable examples, combining multi-modal sensor data, combining sequential data, etc.). There have been some attempts to extended SVM to give probabilistic outputs (see [21], for example). However, this method needs to train another mapping to the SVMs output after the training based on parametric sigmoid mapping. This makes the method more complicated and possibly another validation data set needs to be optimized for post-processing mapping.

To tackle these problems, we apply a Bayesian approach to kernel-based learning via Gaussian process priors. We use the multi-class approach presented in [22], which approximates a complex posterior probability by maximizing the variational lower bound. By using a multinomial probit likelihood model, it is possible to derive a full multiclass classifier as a combination of multiple regression models. These regression models are coupled via the posterior mean estimates of another set of auxiliary variables, which gives a statistically dependent multi-class model. Add-hoc post-processing is not needed. In addition, predictive distribution over unknown examples provides direct confidence measurement as a conditional posterior probability of class labels.

During the training phase of the classifier, Gaussian processes provide a possibility to optimize the hyperparameters by maximizing the marginal likelihood via gradient-based optimization routines [23], [4] or by setting a prior distribution for the hyperparameters and employing sampling methods such as importance sampling to get posterior expectations [22]. We follow the approach used in [22], which uses exponential distribution and a gamma distribution placed on its mean to form a conjugate pair. Furthermore, by applying a radial basis function (RBF) kernel with individual length scale parameters to each feature dimension, we can determine the importance of each feature when optimizing the hyperparameters (i.e., automatic relevance detection (ARD)). This is used to increase the accuracy of person identification as well as to analyze features in different practical settings. One drawback of GPs is that all the training data are needed in the classification phase. When a large data set is used, some sparse approximation methods need to be applied [22]. In this paper, a full model is used due to its capability of real-time performance in the prototype application.

5   Person Identification Based on Floor Sensors

5.1   Feature Extraction

As in typical pattern recognition systems, we need to extract some higher level features from the raw data to be able to perform accurate identification. The binary switch sensor floor forms a matrix where each sensor tile can be presented as a pixel in the image. This allows us to apply standard image processing techniques to detect footsteps and to extract features. We use two kinds of presentations: binary and grey-level images.

A binary image is detected by summing up the sensor values over time, and then thresholding each positive value to one. The summing is performed over each walking sequence. A binary image gives us a direct way to detect the position of each footstep in a sequence. This is done by labeling the 8-connected components of the image. Furthermore, when collecting each individual image in a sequence, we are able to detect the starting and ending time of each connected component for feature extraction.

In addition, the integrated image (i.e., sequence of summed sensor matrices) is saved without thresholding. This matrix presents a grey-level image in which each pixel forms a duration value over the sequence and provides a possibility to extract a rich set of features from the connected components. A "duration map" is presented as a grey-level image in Figure 2, where a brighter value means more time is spent in that position. A binary image can be calculated by thresholding grey-level sensor values larger than 0 to 1.

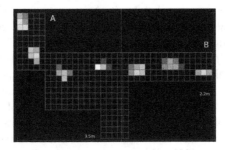

Fig. 2. Grey-level image calculated from sensor measurements of a walking sequence. In addition, the size of the sensor area is illustrated.

Feature extraction is based on the connected components found in the binary image. The features can be divided into two categories: micro and macro-level features. Micro-level features are extracted from each footstep using both the binary and grey-level presentations. This feature set includes features such as the sum of binary pixels in a single footstep profile. Minimum, maximum, mean, and standard deviation values are also extracted from the grey-level component. All these features describe the shape of the "duration map" inside a single footstep profile. To describe the spatial properties of shape, convolution filters, familiar from image processing, are used. We apply four different 3x3 line detection filters and four different 3x3 sobel gradient filters (see, for example, [24] for details). After filtering, the values inside the connected components are summed. Also, the length and width of the footstep, the compensated center of masses, and the duration of the footstep are calculated. Macro-level features present useful information between consecutive footsteps. We use Euclidean distances between the center of mass points of adjacent footsteps as well as individual distances in the longitudinal and transversal walking directions. They are closely related to step length measurement used in gait analysis. Finally, the duration between the starting times of consecutive footsteps is calculated. Macro features are always calculated against the previous footstep in a sequence. A total of 28 features were extracted and are presented in Table 1. It is also straightforward to modify the footstep detection and feature extraction techniques for a real-time application, which is discussed in section 7.

Table 1. Spatial, statistical, and time-related features derived from each footstep profile (1-20) as well as between consecutive footstep profiles (21-28)

Number	Name	Description
1.	sum_{bin}	Number of activated pixels (i.e. sensor tiles) in this footstep profile
2.	sum_{grey}	Sum of grey-level pixel values
3.	min_{grey}	Minimum grey-level value
4.	max_{grey}	maximum grey-level value
5.	$mean_{grey}$	Mean of grey-level pixels
6.	std_{grey}	Standard deviation of grey-level pixels
7.	sum_{vline}	Sum of grey-level component filtered with 3x3 line mask (vertical)
8.	sum_{hline}	Sum of grey-level component filtered with 3x3 line mask (horizontal)
9.	sum_{lline}	Sum of grey-level component filtered with 3x3 line mask (left diagonal)
10.	sum_{rline}	Sum of grey-level component filtered with 3x3 line mask (right diagonal)
11.	sum_{bgrad}	Sum of grey-level component filtered with 3x3 gradient mask (ball of the footstep)
12.	sum_{rgrad}	Sum of grey-level component filtered with 3x3 gradient mask (right side of the footstep)
13.	sum_{hgrad}	Sum of grey-level component filtered with 3x3 gradient mask (heel of the footstep)
14.	sum_{lgrad}	Sum of grey-level component filtered with 3x3 gradient mask (left side of the footstep)
15.	$length_{bin}$	Maximum length of connected binary pixels (longitudinal direction of walking)
16.	$width_{bin}$	Maximum width of connected binary pixels (transversal direction of walking)
17.	com_{bin_x}	Center of mass of connected binary pixels (longitudinal direction of walking)
18.	com_{bin_y}	Center of mass of connected binary pixels (transversal direction of walking)
19.	com_{grey_x}	Center of mass of connected grey-level pixels (longitudinal direction of walking))
20.	com_{grey_y}	Center of mass of connected grey-level pixels (transversal direction of walking)
21.	$duration_{inside}$	Duration of footstep (i.e., activated tiles over time)
22.	$distance_{bin}$	Euclidean distance from previous footstep (using binary center of mass)
23.	$distance_{grey}$	Euclidean distance from previous footstep (using grey-level center of mass)
24.	$duration_{between}$	Duration from the previous footstep (to beginning time of this footstep in milliseconds)
25.	$distance_{bin_x}$	Longitudinal distance from previous footstep (using binary center of mass)
26.	$distance_{bin_y}$	Transversal distance from previous footstep (using binary center of mass)
27.	$distance_{grey_x}$	Longitudinal distance from previous footstep (using grey-level center of mass)
28.	$distance_{grey_y}$	Transversal distance from previous footstep (using grey-level center of mass)

5.2   Person Identification: Single Footsteps and Walking Sequences

We derive two kinds of person identification methodologies based on the multi-class Gaussian process classification and features presented in the previous section. The first one is a conventional classification scenario where we use posterior distribution of class labels predicted from a single footstep profile to make the decision. In this case we use micro-features as well as macro-features related to the previous footstep. This scenario is useful in situations where the decision has to be made as quickly as possible.

On the other hand, if we want more accurate recognition, we can use classification information from multiple adjacent footstep profiles by combining the posterior distribution of class labels. This scenario gives a recognition based on a sequence, which in this case is one walking sequence (5-7 footsteps) on the floor. As GP classification provides posterior over class labels, we can use summation and product rules to combine the outputs. This kind of rule has been shown to be simple, yet powerful, in many information fusion problems [25]. The advantage is that we can use a conventional training phase and an arbitrary number of examples in a sequence to make the final decision. If $P(\omega_k|x_i)$ represents the posterior probability of class labels $(1 \ldots n)$ conditioned on unknown example x_i and S is the total length of a sequence, the final decision can be calculated using the sum (Eq. 1) and product rule (Eq. 2), as follows:

$$\omega_c = \underset{k=1}{\overset{n}{\mathrm{argmax}}} \left[\sum_{i=1}^{S} P(\omega_k|x_i) \right] \tag{1}$$

$$\omega_c = \underset{k=1}{\overset{n}{\mathrm{argmax}}} \left[\prod_{i=1}^{S} P(\omega_k|x_i) \right] \tag{2}$$

The disadvantages of using this kind of scenario are related to optimization. Due to the fact that the model is trained on single footsteps, it does not use information of sequences to find a global optimum. In addition, the choice of combination rules in our scenario is more ad-hoc and experimental compared with approaches where sequential information is directly learned from the data. However, this simple approach is able to use the information of walking sequences at some level to be able to produce more accurate decisions, as is shown in the results section. A comparison with more advanced models, such as sequential kernels and other sequential classifiers, is left for future work.

6   Results

6.1   Data Sets

To test the identification methods presented here, we collected a large data set. The data set included walking sequences of nine different subjects. The test group consisted of two female and seven male subjects, and each wore their own shoes (which were indoor sandals in this case). They were told to walk their natural walking speed over the sensor floor (from A to B in Figure 2) 20 times. To get as natural a data set as possible, the starting foot or the absolute position of each footstep in the sequence was not constrained in any way. Each sequence included 5-7 footstep profiles, depending on the stride length of the subject. Altogether 1143 footstep profiles were collected from the nine walkers.

In addition, to examine the effect of different walking styles (i.e., walking speed) and footwear on identification, we collected more data from four subjects. To study variations in walking speed, we recorded additional sequences in which the subjects were told to walk slower and faster than usual. Both settings were performed 10 times. To test the effect of different footwear, 20 sequences of subjects wearing their own outdoor trackers and no shoes at all were collected. Combining this data set with the footsteps of the four persons collected earlier gave us 1981 footstep profiles for studying the effect of variation in walking speed and footwear.

A total of 2597 footstep profiles were collected in these sessions. To test and analyze the usefulness of the features and the classification method as well as the modeling capability of the features and adaptation of the classifier to novel data, we split the data set into different subgroups. The standard nine-person data set included 20 sequences of normal walking speed and sandals for studying the extracted features and the capability to perform multi-class classification using Gaussian processes. To analyze the effects of variations on the extracted features more precisely, the footstep profiles of four persons were divided into three subgroups: standard (including walking at normal speed and with sandals), footwear (including three different footwear at normal speed), speed (including three different speeds with sandals on). The aim of these data sets was to be able test how well the extracted features can handle variations in the data set and which features have the best discriminative power in these settings.

Furthermore, we split the four-person data set into 12 subgroups: sandals (including all the data from sandals), without sandals (all the data except from sandals), trackers (including data from outdoor shoes), without trackers (including all the data except from trackers), without shoes (including the session without shoes), shoes (including the session with shoes), normal (including normal speed), not normal (including slow and fast walking), slow (including slow walking), not slow (including normal and fast walking), fast (including fast walking), not fast (including slow and normal walking). These data sets were used to examine the generalization capability of the classifier and the need for adaptation when the test data set includes differently distributed (in this case walking speed and footwear) data. These are very important when building practical applications. A summary of the data set categories is presented in Table 2.

Table 2. Summary of different data set categories used in the person identification experiments

Number	Name	Description	Number of examples	number of sequences
1.	9 persons standard	Normal walking speed with sandals	1143	180
2.	4 persons standard	Normal walking speed with sandals	527	80
3.	Footwear	Normal walking speed with footwear variations	1516	240
4.	Speed	Slow, normal, and fast walking speed with sandals	992	160
5.	Sandals	All the data with sandals	992	160
6.	Without sandals	All the data without sandals	989	160
7.	Trackers	All the data with trackers	441	80
8.	Without Trackers	All the data without trackers	1540	240
9.	Shoes	all the data with shoes	1433	240
10.	Without Shoes	All the data without shoes	548	80
11.	Normal	All the data with normal speed	1516	240
12.	Without normal	All the data without normal speed	465	80
13.	Slow	All the data with slow speed	248	40
14.	Without slow	All the data without slow speed	744	180
15.	Fast	All the data with fast speed	215	40
16.	Without fast	All the data without fast speed	755	180

6.2 Person Identification

In this section we present the recognition result of using the nine-subject data set described in Section 6.1. We split the data set so that 2/3 were used for training and 1/3 for testing, and all the features were scaled between 0 and 1. Variational GP approximation was achieved using 10 iterations, simultaneously learning the hyperparameters of the RBF kernel [20], [22]. This was repeated 10 times on randomly chosen training and test sets. All the tests were implemented with Python programming language and the GP models were trained with an R language variational Bayesian GP package [26].

Furthermore, sequential recognition was tested by combining the GP outputs using similarly trained models and fixed sum and product rules. Table 3 presents the average total identification (and standard deviations) of single footstep profiles as well as combined recognition rates. The classifier is able to classify correctly 64% of the individual footsteps, which shows the complexity of the data set obtained from the simple binary switch sensors. Using the fixed combination rules increases accuracy and the product rule outperforms the sum rule in this data set, showing an 84% success rate. The results show that to achieve a high success rate, sequential information is needed.

Table 3. Total identification accuracies of recognizing nine different walkers

	GP (single examples)	GP (sum rule)	GP (product rule)
Accuracy (%)	64.23 (3.27)	82.33 (6.59)	84.26 (6.69)

6.3   Feature Analysis of Footwear and Walking Speed Variations

This section presents the results of analyzing the effect of different footwear and walking speed variations. Moreover, we rank the individual features based on their relevance in the identification method to determine which are the best and worst ones. To our knowledge, this is the first time both footwear and walking speed changes are analyzed in the context of floor sensors. These are very important issues when building a practical identification system.

We used the different four-person data sets presented in Table 2, where we summarize the total success rates (accuracy) as well as the most relevant features (mrf) and least relevant features (lrf) (cf. Table 1 for the order number of the features). Table 4 presents the results using standard data sets and footwear/speed variations. Looking at the accuracies, the total number of persons in a classification has a large impact (nine persons vs. four persons.). Secondly, footwear variation slightly decreases accuracy compared with the standard data set (4.36 percent units). Walking speed decreases accuracy much more (10.50 percent units). In all the data sets, the most important features are related to walking sequence (i.e., $distance_{bin}$, $distance_{grey}$, $duration_{between}$) and the duration of footsteps. The least relevant features change, but are always related to micro-features. These results indicate that when using limited binary sensors, the use of features carrying sequential information is very important. The average length scales of each feature in the nine-person data set are presented in Figure 3. A smaller value means the feature is more important in the classification decision. The walking sequence features are the most important, but footstep shape features (e.g., calculated by the convolution filters) have a large impact, too (e.g., features 8, 10 and 14)

Similar experiments are shown in Table 5. Now the test set contains variations (i.e., footwear and walking speed) that are not included in the training data set. This is the most complex approach presented in the paper. Clearly, a large decrease in total accuracies can be seen when comparing the results with those in Table 4. This indicates that it is important to collect and use all available information for training if these variations are assumed to happen. Similarly, it can be concluded that speed variations have a larger negative impact on accuracy compared with footwear variations. Interestingly, the same features as in the above data sets have the most relevant information for identification, on average.

Table 4. Total identification accuracies and feature ranking using different datasets. The data sets are described in Table 2 and the features are presented in Table 1. The three most relevant features (mrf) and least relevant features (lrf) are shown.

Dataset	Accuracy (%)	mrf	lrf
9 persons standard (1.)	64.23 (3.27)	21.,24.,23.	2.,28.,20.
4 persons standard (2.)	81.45 (1.62)	21.,23.,24.	16.,20.,3.
Footwear (3.)	77.09 (1.22)	24.,21.,22.	12.,11.,4.
Speed (4.)	70.95 (2.20)	21.,23.,24.	3.,19.,20.

7   Prototype Application: Context-Aware Reminder

A prototype application was built based on single footstep identification. A multi-class Gaussian process classifier was learned from the training data set of four laboratory

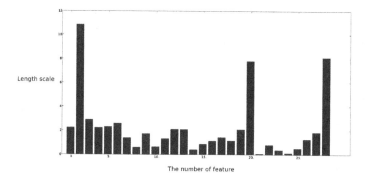

Fig. 3. RBF kernel length scales of each feature using a nine-persons data set. The horizontal axis presents the feature number from Table 1 and and the vertical axis describes the importance of the feature, where a smaller length scale value means the feature is more important.

Table 5. Total identification accuracies and feature ranking using different data sets. The data sets are described in Table 2 and features are presented in Table 1. The three most relevant features (mrf) and least relevant features (lrf) are shown.

Train	Test	Accuracy (%)	mrf	lrf
Without sandals (6.)	Sandals (5.)	59.68	24.,23.,21.	15.,13.,5.
Without trackers (8.)	trackers (7.)	59.49	23.,24.,21.	13.,16.,9.
Shoes (9.)	Without shoes (10.)	59.85	21.,24.,23.	26.,1.,14.
Normal speed (11.)	Without normal (12.)	48.60	21.,27.,24.	5.,7.,3.
Without slow (14.)	Slow speed (13.)	57.66	21.,23.,24.	5.,3. 12.
Without fast (16.)	Fast (15.)	41.01	21.,23.,24.	1.,20.,11.

members. In addition, the position of each footstep was calculated using the center of mass in the binary image. This very simple method is able to locate one person at a time. In the future, more advanced tracking methods will be applied to detect the positions of multiple simultaneous walkers.

The prototype was implemented as a distributed system consisting of three different levels, were each level provides information via TCP/IP socket communication. The TCP/IP-based approach was chosen to leverage existing libraries for rapid prototyping, which requires language independence. The first level provides raw sensor data, which is read by the identification system on the second level. The lower-level implementation consists of a Windows DLL (VC++) for InfoFloor driver and Java TCP/IP server software. The identification system extracts features from the raw data and sets the identity based on GP as well as position and the time stamp information of each example. In this application feature extraction needs to be implemented in real time. This was done by monitoring the starting and ending times of connected binary events on the floor, and when these were detected micro-level features were calculated from the sensor area of the footstep using both binary and grey-level presentation. After that macro-level features were calculated based on the detection information from the previous footstep. If a certain time period (e.g., 5 seconds) expired without any events, it was assumed that the person has left the sensor area and the detection phase is starting over again. The second level was implemented with Python language, as presented in the results section. The recognition software worked in real time and it took no more than 20 ms to process the raw data into identification prediction (using the model trained on four-person data). The time between two adjacent footsteps was approximately 500ms. The third level is an application that reads identified events from the identification system. Along with side information about the context of the environment, it provides reminders to a user. The client program was implemented with Java. The components of the software architecture are presented in Figure 4(a). In this application scenario the user interface is implemented with two displays. The first one is located above the refrigerator and the second one is located near the entrance to a "smart room" (see Figures 4(b) and 4(c)). The scenario, which assumes side information, is as follows:

1. *Nobu bought a bottle of milk a week ago and put it into the refrigerator. One week later, when he is passing in front of the refrigerator, it notifies him of the expiring status of the milk. Here, a mirror display is installed on the fridge, and the fridge is capable of determining the status of the contents.*
2. *Nobu, a Tokyo resident, is going on a trip to Kyoto. Although the weather is fine in Tokyo, the weather forecast says it will be rainy in Kyoto. The "smart room" knows his schedule, i.e. date and location, as well as the identify of the person and the walking direction. When he is leaving the room, a display installed at the entrance recommends him to take an umbrella with him because of the forecast.*

This prototype application shows a simple approach to using naturally obtained person identification information, recognized from walking (along with the side information), in a context-aware system.

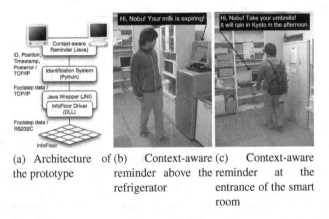

(a) Architecture of the prototype

(b) Context-aware reminder above the refrigerator

(c) Context-aware reminder at the entrance of the smart room

Fig. 4. Software architecture and scenarios in the prototype application

8 Conclusions

In this paper we presented a floor sensor system based on binary switches as well as methods for recognizing a persons identity based on sensor measurements collected from the floor. In addition we showed a prototype application that uses the information of a walkers identity and the position of footsteps to provide context-aware reminders for daily life. For the recognition purposes, a set of useful features were extracted from the raw measurements. The measurements are presented as binary and grey-level images, which allow us to use basic image processing methods to derive higher-level features. A variational Bayesian approximation of a multi-class Gaussian process (GP) classifier is used to identify the walkers. As a Bayesian method the GP gives the posterior distribution of predicted class labels. This information was used to combine the classifier outputs of multiple footsteps using conventional classifier combination rules. This provides a simple approach to recognizing a sequence of walking in an application where a more accurate decision is needed. The total recognition rates of nine different subjects using individual footsteps as well as walking sequences were 64% and 84%, respectively. This is a very promising result using simple binary switch sensors.

Furthermore, GPs provide a flexible solution to model selection (e.g., the choice of hyperparameters). We used a kernel that is able to weigh each feature's dimensions differently through hyperparameters. This provides automatic relevance detection (ARD), where the most important features get more weigh in a similarity measurement. ARD was used to train an accurate model and to analyze the importance of individual features. We analyzed the effect of different footwear and variations in walking speed on identification accuracy. This kind of analysis is missing from most of the previous studies using floor sensors. In our experiments we found that both of these variations have an impact; walking speed variations have a larger negative impact. Moreover, the most relevant features in all the tested data sets were related to distance and duration between footsteps as well as the duration of a single footstep profile.

Acknowledgments

This work was supported by The Ministry of Education, Culture, Sports, Science and Technology in Japan under a Grant-in-Aid for Division of Young Researchers, and InfoTech Oulu Graduate School.

References

1. Essa, I.A.: Ubiquitous sensing for smart and aware environments: Technologies towards the building of an aware home. IEEE Personal Communications, Special issue on networking the physical world, 47–49 (October 2000)
2. Brummit, B., Meyers, B., Krumm, J., Kern, A., Shafer, S.: Easyliving: Technologies for intelligent environments. In: 2nd International Symposium of Handheld and Ubiquitous Computing (HUC), New York, USA, pp. 12–29. Springer, Heidelberg (2000)
3. Pentland, A.: Smart rooms. Scientific American 274, 68–76 (1996)
4. Rasmussen, C.E., Williams, C.K.I.: Machine Learning for Gaussian Processes. MIT Press, USA (2006)
5. Orr, R.J., Abowd, G.D.: The smart floor: A mechanism for natural user identification and tracking. In: Proceedings of 2000 Conf. Human Factors in Computing Systems (CHI), The Hague, Netherlands, pp. 275–276. ACM Press, New York (2000)
6. Addlesee, M.D., Jones, A., Livesey, F., Samaria, F.: ORL active floor. IEEE Personal Communications 4(5), 35–41 (1997)
7. Cattin, P.: Biometric Authentication System Using Human Gait. PhD thesis, ETH-Zürich, Institute of Robotics, Switzerland (2002)

8. Rodriguez, R.V., Lewis, R.P., Mason, J.S.D., Evans, N.W.D.: Footstep recognition for a samrt home environment. International Journal of Smart Home 2(2), 95–110 (2008)

9. Suutala, J., Röning, J.: Methods for person identification on a pressure-sensitive floor: Experiments with multiple classifiers and reject option. Information Fusion Journal, Special Issue on Applications of Ensemble Methods 9, 21–40 (2008)

10. Yun, J.-S., Lee, S.-H., Woo, W.-T., Ryu, J.-H.: The user identification system using walking pattern over the ubifloor. In: Proceedings of International Conference on Control, Automation, and Systems (ICCAS), Gyeongju, Korea (October 2003)

11. Middleton, L., Buss, A.A., Bazin, A., Nixon, M.S.: A floor sensor system for gait recognition. In: Fourth IEEE Workshop on Automatic Identification Advanced Technologies (AutoID 2005), pp. 171–176 (2005)

12. Murakita, T., Ikeda, T., Ishiguro, H.: Human tracking using floor sensors based on the markov chain monte carlo method. In: Proceedings of Seventeenth International Conference on Pattern Recognition (ICPR), Cambridge, UK, August 2004, pp. 917–920 (2004)

13. Morishita, H., Fukui, R., Sato, T.: High resolution pressure sensor distributed floor for future human-robot symbiosis environment. In: Proceedings of the IEEE/RSJ International Conference on Intelligent Robots and Systems, EPFL, Lausanne, Switzerland, pp. 1246–1251 (October 2002)

14. Sudo, K., Yamato, J., Tomono, A.: Determining gender of walking people using multiple sensors. In: Proceedings of the IEEE/SICE/RSJ International Conference on Multisensor Fusion and Integration for Intelligent Systems, December 8-11, 1996, pp. 641–646 (1996)

15. Eng, K., Douglas, R.J., Verschure, P.F.M.J.: An interactive space that learns to influence of human behaviour. IEEE Transaction on Systems, Man, and Cybernetics-Part A: Systems and Humans 35(1), 66–77 (2005)

16. Headon, R., Curwen, R.: Recognizing movements from the ground reaction force. In: Proceedings of the Workshop on Perceptive User Interfaces, Orlando, Florida, USA, November 15-16, 2001, pp. 1–8 (2001)

17. Köhle, M., Merk, D.: Clinical gait analysis by neural networks: Issues and experiences. In: Proceedings of the IEEE Symposium on Computer-Based Medical Systems, Maribor, Slovenia, pp. 138–143 (1997)

18. Vstone corporation, http://www.vstone.co.jp/e/etop.html

19. Cristianini, N., Shawe-Taylor, J.: An Introduction to Support Vector Machines and Other Kernel-based Learning Methods. Cambridge University Press, UK (2000)

20. Schölkopf, B., Smola, A.: Learning with Kernels: Support Vector Machines, Regularization, Optimization, and Beyond. MIT Press, USA (2001)

21. Platt, J.: Probabilistic outputs for support vector machines and comparisons to regularized likelihood methods. In: Smola, A., Bartlett, P., Schölkopf, B., Schuurmans, D. (eds.) Advances in Kernel Methods - Support Vector Learning, pp. 61–74. MIT Press, Cambridge (1999)

22. Girolami, M., Rogers, S.: Variational bayesian multinomial probit regression with gaussian process priors. Neural Computation 18, 1790–1817 (2006)

23. MacKay, D.J.C.: Information Theory, Inference, and Learning Algorithms. Cambridge University Press, UK (2003)

24. Gonzalez, R.C., Woods, R.E.: Digital Image Processing. Prentice-Hall, Englewood Cliffs (2002)

25. Kittler, J., Hatef, M., Duin, R.P.W., Matas, J.: On combining classifiers. IEEE Transactions on Pattern Analysis and Machine Intelligence 20(3), 226–239 (1998)

26. Lama, N., Girolami, M.: Vbmp: Variational bayesian multinomial probit regression for multiclass classification in R. Bioinformatics (2008)

GammaSense: Infrastructureless Positioning Using Background Radioactivity

Doina Bucur and Mikkel Baun Kjærgaard

Department of Computer Science, University of Aarhus, Denmark
{doina,mikkelbk}@daimi.au.dk

Abstract. We introduce the harvesting of natural background radioactivity for positioning. Using a standard Geiger-Müller counter as sensor, we fingerprint the natural levels of gamma radiation with the aim of then roughly pinpointing the position of a client in terms of interfloor, intrafloor, and indoor-versus-outdoor locations. We find that the performance of a machine-learning algorithm in detecting position varies with the building, and is highest for interfloor detection in the case of an old domestic house, while it is highest for intrafloor detection if the floor spans building segments made from different construction materials. Altogether, the technique has lower performance than infrastructure-based localization techniques.

1   Introduction

Positioning is an important requirement for many novel applications. However, technologies for positioning often depend on extensive infrastructures, which limit the coverage of these technologies and create breakdowns in the user experience when a user crosses the—often invisible—infrastructure boundaries. One approach for solving this problem is the development of positioning technologies with pervasive coverage which minimizes the dependence on infrastructure.

Prior work on positioning has used the properties of physical phenomena such as hearable sound, ultrasound, and types of non-ionizing radiation (e.g. visible light and radio signals). In this work we consider the use of ionizing radiation (specifically, products of radioactivity: alpha, beta and gamma rays) for positioning. In our environment, the sources of such radiation include natural radioactivity in the soil and building materials, and cosmic rays.

In practice, gamma radiation is most relevant, since alpha and beta radiation have subcentimeter propagation range. The radiation appears naturally in the environment, and the geometry of its sources in a built environment gives more interesting variation patterns than the patterns of other radiation sources, such as visible light.

Harvesting gamma radiation to infer location has several advantages: (i) Gamma radiation is pervasive, therefore coverage is not confined to an area of infrastructure coverage; (ii) The geometry of radiation sources is strikingly different from that of other radiation, which makes gamma readings a desirable sensor input

D. Roggen et al. (Eds.): EuroSSC 2008, LNCS 5279, pp. 69–82, 2008.

in e. g. sensor fusion; (iii) There exist many devices that sense gamma radiation, and the threat of terror might make them even more widespread. Doing positioning over gamma radiation also has disadvantages: (i) Radiation has no identifier embedded in the signal, so there exists only one signal source to use in location estimation. This is the same drawback as when using visible light. (ii) Compared to other signals, a longer sampling time is needed for detecting relevant statistical properties of received radiation patterns.

We make the following contributions: (i) We show that gamma radiation is a predictable signal for positioning (ii) We analyze which environment properties gamma radiation depends on (iii) We present the first positioning system based on background radioactivity named *GammaSense*, which positions a device using location fingerprinting over gamma readings.

The remainder of this paper is structured as follows: In Section 2, we present the relevant related work. Subsequently, we give a primer to gamma radiation and discuss it as a signal source for positioning in Section 3. Then the novel GammaSense positioning system is introduced in Section 4. Evaluation results for GammaSense are provided in Section 5. Finally, Section 6 concludes the paper and provides directions for future work.

2   Related Work

Our system examines the performance of doing localization on the basis of measuring the levels of background radioactivity (gamma rays) indoors, and employing the technique of location fingerprinting. While—to our knowledge—no other work does infrastructureless localization with background radioactivity, a wealth of existing research does explore the matter of doing localization indoors, based on a variety of technologies and types of infrastructure, such as infrared, ultrasonic and ultra-wideband, and their specific transceivers. The common drawback of these systems is their reliance on custom infrastructure, a fact which then diminishes their acceptance and easiness of deployment. GammaSense takes the very opposite approach and looks into making use of natural signals for indoor positioning, offering a degree of location detection completely independent of any infrastructure.

GammaSense uses the technique of *location fingerprinting*, which has already been applied in related work with radio waves, light and sound signals. It is based on the acquiring of a database of prerecorded signal measurements, denoted as location fingerprints. Clients' locations are then estimated by comparing them with the database of fingerprints [1].

One of the first location fingerprinting systems was the radio-based RADAR [2] system, which applied deterministic mathematical models to calculate the co-ordinates of a client's position from WaveLan/IEEE 802.11 signal strengths. Similar methods were applied to GSM signals by Otsason et al. [3]. Unlike RADAR, later systems employed probabilistic models instead of deterministic ones. An example of a probabilistic positioning system was Youssef et al. [4]; a similar one determining the logical position or cell of a client was published by Haeberlen et al. [5]. A radio-based system named SkyFloor [6] focused on predicting the floor of a client.

The principle of location fingerprinting was also applied to sound signals by Patel et al. [7]; their system uses the electric wiring in a building to generate sound signals on several frequencies, which form distinctive sound patterns throughout the building. A tone detector then picks up these sound signals and uses them as location fingerprints. Also, a positioning system based on location fingerprinting over visible light intensities was proposed by Ravi et al. [8].

3 Gamma Radiation

3.1 A Primer on Radioactivity and Ionizing Radiation

Radioactivity is the natural phenomenon in which certain—possibly artificial—chemical elements emit radiation spontaneously, be it in the form of electromagnetic waves or of charged particles. The cause of this emission is radioactive decay, i.e. the spontaneous transmutation of an unstable parent element into a more stable daughter element; the decay rate is practically expressed with the term "half-life", meaning the span of time required for half the quantity of the radioactive element to transmute.

One form of radioactive decay is the beta decay. A neutron or a proton transform into the other within the parent nucleus, accompanied by the emission of an electron (or its positively charged version, a positron); this electron is called a *beta particle* or beta ray. If a neutron transmutes into a proton, a negatively charged electron e^- with high kinetic energy (a β^- particle) is expelled from the nucleus. The alternative transmutation with a positron emission (a β^+ particle) cannot occur without energy input, because the mass of a neutron is higher than that of a proton. Hence, β^+ rays are mostly produced artificially in particle accelerators ([9]).

Another produce of radioactive decay are *gamma rays*, an electromagnetic radiation having the highest frequency and the shortest wavelength within the electromagnetic spectrum. Neutrons and protons occupy well-defined energy levels in a nucleus, and when either particle is excited to a higher unoccupied level, the excited nucleus decays to a lower energy state, and the difference in energy is emitted as gamma radiation. This energy difference is much larger (in the range of MeV) than in the case of excited electrons, whose similar state mutations release visible, near-visible light or X-rays (with an energy of a few eV).

Beta particles typically have energies from a few KeV to a few MeV and the mass of an electron (atomic mass 1/1836). The range of beta particles is short: a 1.9mm sheet of aluminium stops a 1MeV beta particle ([10]). Because they are relatively light, beta particles do not travel in straight lines but follow a random path through material. Gamma rays are high-energy (0.1 to 3 MeV). Their range is long and penetration power high; their typical means of losing energy is by ejecting an electron from an atom and being scattered from the impact with reduced energy (the Compton effect). To reduce the intensity of a

0.5MeV gamma ray to 0.37 of its initial value, one needs to lay a shield of 4cm of aluminium, 0.59cm of lead or 28.6cm of tissue paper ([10]).

Beta and gamma radiation also accompany *alpha decay* (another omnipresent radioactive decay in which heavy nuclei disintegrate by emitting a positively charged nucleus of helium, $_2^4\text{He}^{2+}$, with an even shorter range than beta particles, due to their large atomic mass of 4). Many alpha sources are accompanied by beta-emitting radiodaughters, and alpha emission is followed by gamma emission from the remaining negatively charged nucleus.

All three forms of radiation produced by radioactive decay can ionize atoms or molecules in their path (either due to their electric charge, or to their kinetic energy), transforming them into ions by adding or removing changed particles. Hence, along with other rays, they are collectively called *ionizing radiation*.

3.2   Natural Sources of Radioactivity

Background radiation is omnipresent, and has always existed naturally. Its sources are *cosmic rays* (from outer space and the Sun) and *terrestrial radioactivity* naturally occurring in soil, building materials and in air, water, foods and the human body (as in Table 1, after [11]).

Table 1. Average worldwide exposure to natural air and soil radiation sources

Source of radiation	Percentage
Cosmic radiation total	18.48
of which ionizing (beta and gamma)	13.74
Terrestrial radiation total	82.46
soil and building materials: ^{238}U, ^{232}Th (alpha), ^{40}K (beta)	22.74
air: radon ^{222}Rn, thoron ^{220}Rn (alpha)	59.24

Out of the sources of radiation in Table 1, the ionizing component of the cosmic radiation, the radioactive elements in soil and construction materials (^{238}U, ^{232}Th series and ^{40}K, in approximately equal contributions), and the aerial radioactive gases radon (and, in small concentrations, thoron) are all measurable sources of beta and gamma rays (either directly or indirectly, as a side effect of alpha decay).

The specific concentrations of radioactive elements in soil are related to the types of rock from which the soils originate, which in turn correlate with the concentrations in air. The gas radon (a decay product of radium, with a half-life of 3.8 days) diffuses out of the soil. Radon and its decay products are the most important contributors to human exposure to radiation from natural sources.

Indoor concentration of gamma rays, mainly determined by the materials of construction, is inherently greater than outdoor exposure if earth materials have been used; the geometry of the radiation source changes indoors from a half-space to a surrounding configuration. The indoor to outdoor ratios range from 0.6 to 2.3, with a worldwide ratio of 1.4 ([11]).

3.3   Indoor Variations of Radiation Concentrations

Some of the radioactive sources are fairly constant and uniform geographically, while others vary widely with location. Naturally, cosmic rays are less intense at lower altitudes, and concentrations of uranium and thorium are elevated in certain soils. More interestingly, the building materials of houses and the design and ventilation systems strongly influence indoor levels of the most important contributors, radon and its decay products.

Advection from the soil is the main factor for high radon entry rates in buildings (as in Table 2, after [11,12]). Radon is driven indoors by the pressure differential between the building and the ground around the foundation, produced by the higher indoor temperatures, ventilation, and to some degree by wind blowing on the building. The effectiveness of this pressure differential is dependent on the permeabilities of the building foundation and the adjacent earth. Also, wind can cause decreases in radon entry concentrations by its flushing effect on radon in soil surrounding the house. Because of differences in the pressure differentials and permeabilities, advection varies greatly from structure to structure, especially in temperate and cold climates. The non-masonry building in Table 2 has less accumulation of radon than the masonry one, due to a smaller radiation contribution from walls and ceilings.

Table 2. Representative radon entry rates in low-level residential houses: a masonry house and a wooden house in Finland [11]

Source of radon	Mechanism	Rate (Bq/m^3h) and percentage	
		Masonry house	Wooden house
Building elements			
Walls and ceiling	Diffusion	16 (18)	2 (3)
Subjacent earth			
Through gaps	Advection	66 (73)	60 (86)
Through slab	Diffusion	4 (4)	4 (6)
Outdoor air	Infiltration	3 (3)	3 (4)
Water supply	De-emanation	1 (1)	1 (1)
Total		90 (100)	70 (100)

Various surveys also find radon concentration variations between rooms in the same building. Ghany [13] and Sonkawade et al. [14] find that the mean values of radon concentration in bathrooms and kitchens were significantly higher than those in living rooms and bedrooms. The find is motivated by ceramic being a radon source, poor ventilation and the use of underground water and natural gas.

3.4   Experimental Results

Given the representative surveys in Subsections 3.2 and 3.3 upon the variations of radioactivity levels indoors, we expect to be able to harvest indoor radiation levels to use in localization. Two facts about the variability of the radiation levels are important when designing a localization algorithms based on fingerprinting: one is the geometric variability of the measurements within the building, and the other is the consistency of the signal levels at any given location. We give a brief account of our findings in the following.

Our experiments did confirm the expected indoor geometry of the radioactivity source. In one of our testbed buildings (a two-level domestic house, as seen later in Subsection 5.1), both the mean and the variance of radioactivity readings differ visibly between the two rooms on different floors, as in Fig. 1. The mean over a 1-hour-long continuous length of 1-minute counts shows a 10.32% decrease from the ground-floor room (mean 30.79 counts per minute) to the first-floor room (mean 27.61 counts per minute), and a striking difference in standard deviations (a 26.43% decrease, from 5.75 on level zero to 4.23 on the first level). This fact verifies that the major radiation source indoors is the subjacent earth, with its influence being diminished and smoothed with rising levels.

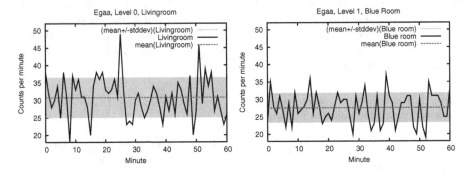

Fig. 1. The difference in radiation measurements over 1-hour periods between rooms on different floors in a domestic house in Egaa, Denmark

To confirm the stability of the signal, we then refer to Fig. 2, which depicts the typical variation in readings in a fixed position over a 1-day-long period of time. In another of our testbeds (a public institution, more on which in Subsection 5.1), subjected to a fair level of human activity during a working day, the signal doesn't exhibit significant variation.

4   GammaSense

The sensor we used for measuring radiation levels was a radioactivity meter composed of two devices: a commercially available Geiger-Müller tube and a pulse ratemeter (i. e. counter, either battery- or AC-powered) designed to feed the Geiger-Müller tube with voltage and to handle the pulses delivered by the tube. Both were manufactured by a small local company called *Impo electronic*, for use in education. The setup and the tube's technical specification are depicted in Fig. 3.

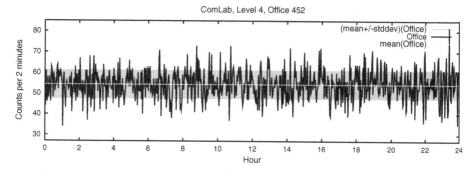

Fig. 2. Signal stability over a 24-hour period

A Geiger-Müller tube is one of the several radioactivity detectors whose functionality is based on the ionizing capability of the radiation produced through radioactive decay. The tube is filled with a mixture of inert gases, in which incoming ionizing radiation creates electrons and positively charged ions. The tube wall constitutes the negative electrode or cathode, while a thin central wire is highly charged with positive voltage and is an anode. The strong electric field created by the electrodes accelerates the ion towards the cathode and the electron towards the anode, giving them sufficient energy to ionize further gas molecules through collisions, thus creating an avalanche of charged particles.

The result is a short, intense pulse of current which passes from the negative electrode to the positive electrode and is counted as one ionizing particle. The counter also includes an audio amplifier that produces a beep upon pulse detection. The number of pulses per time unit measures the intensity of the radiation field. Cheap and robust, a Geiger-Müller tube can detect the intensity of radiation (particle frequency), but not particle energy.

We performed all measurements maintaining the anode voltage strictly at the recommended, mid-plateau voltage of 575V (as described in Fig. 3(b)), for best independence of counts from voltage supplied, both when powering the device from battery and when plugged into the power supply infrastructure. The counter allows the configuration of a small number of counting parameters, such as different count times (between 1 and 120 seconds) after which the total number of particles is reported, and has the ability to count and report continuously for such subsequent fixed periods.

The results are either reported on the counter's small screen, or sent through the counter's RS-232 interface to a laptop, stored in a volatile memory (which only holds 50 counts), or stored in a non-volatile datalog (with a capacity of 250 counts). In our experiments, we used either the sending of each count in real time to the laptop, or the downloading of the contents of the datalog—when full—to the laptop; in both cases, the readings were saved in the ratemeter's format in raw data files.

On the laptop side of the RS-232 interface, we used a Linux Debian with *minicom*, a text-based, GPL-licensed terminal emulator for Unix-based operating systems.

Sensitive to:	beta, gamma rays.
Effective diameter:	27.8mm.
Window material:	mica.
Window thickness:	$[2\frac{mg}{cm^2}, 3\frac{mg}{cm^2}]$.
Gas filling:	neon, argon, halogen.
Plateau:	$[450V, 700V]$.
Recommended voltage:	575V.
Dead time at 575V:	$19\mu S$.

(a) Geiger-Müller tube in holder (left) and pulse ratemeter with data log and serial interface (right)

(b) The tube's technical specifications. The *plateau* is the voltage interval where counts/sec is least dependent of voltage.

Fig. 3. The experimental setup: Geiger-Müller tube and counter with technical specifications

For implementing location fingerprinting using gamma readings we used the machine-learning tool Weka [15]. Weka implements a range of machine-learning algorithms and several GUIs for configuration, experimentation, and visualization. Before selecting a machine-learning algorithm to use, we experimented with different algorithms, and concluded that the *LogitBoost* algorithm was most fit. LogitBoost is based on additive logistic regression, which means that the algorithm—given a simpler algorithm—iteratively constructs a better detector by combining several instances of the simple algorithm. We used a decision stump (i.e. a simple boolean classifier) as the simple algorithm.

In order to feed our data to Weka, we implemented a small Java program that reads the raw data files and outputs ARFF data for the Weka tool. The program also calculates certain features from the raw gamma readings: aggregated 60-second counts, and the mean and variance of the latest five 60-second counts. These features are then written to the ARFF file together with ground truth about whether the data was collected indoor or outdoor, on which floor, and at which horizontal location.

5   Evaluation

5.1   Data Collection

We collected the location fingerprints we used in our indoor localization study over an 8-month period ending in June 2008. For the study, we chose four testbed buildings with diverse construction parameters (i. e. number of floors, building materials and age of construction) located in two countries. We give a superficial visual impression of the four testbed buildings in Fig. 4, where we also list the identifiers by which we refer to the testbeds in the rest of the paper.

Their construction parameters are then recorded in Table 3, in terms of the year the building was finished, the type of construction materials and the number

(a) an old village house in Egaa, Denmark (Egaa)

(b) the Computing Laboratory at the University of Oxford, UK (ComLab)

(c) the Mathematics Department at the University of Aarhus, Denmark (Math)

(d) the Computer Science Department at the University of Aarhus, Denmark (CS)

Fig. 4. The testbed buildings, ordered by age of construction; the building identifier is listed in parentheses

of building levels. For building Egaa, 1829 is the building year, while 1960 is the year when three of the perimeter walls have been renewed. Testbed ComLab is an extensive building composed of a number of smaller constructions, linked together: four old Victorian houses built before 1901 formed the southern wing, to which a identically-styled northern wing was added by 1993, for then a fully-modern segment to be attached by 2006.

Using the setup in Fig. 3 detailed in Section 4, we collected fingerprints of the background radioactivity at fixed height for any particular building. For Egaa, we took sample counts in each of the four rooms for one hour, while for CS and Math we fixed one and three locations, respectively, on each floor, and took sample counts for 10 to 20 minutes. In ComLab, due to the complexity of the building, we divided the study cases in two: a vertical study aiming at distinguishing among floors collected 20-minutes worth of samples from the same vertical location on each level, while for the horizontal study we took measurements on the fourth floor at six locations divided equally among the three wings of the building.

For an additional indoor-versus-outdoor study, in the case of ComLab we also collected samples from two outdoor locations in the very vicinity of the testbed

Table 3. The testbeds' characteristics

Building	Built by	Building materials	Floors	Use
Egaa	1829, 1960	Stone, brick, wood, straw	2	domestic
ComLab	1901, 1993, 2006	Brick, concrete	5	institution
Math	1967	Brick, concrete, stone, wood	5	institution
CS	2001	Concrete, steel, glass	4	institution

building. Also, to verify the stability of the signal, we collected long-term counts: one over 5 days in a fixed location in a CS office, and one over 2 days in a ComLab office.

All sample measurements were collected as continuous counts, 10-second (in the majority of experiments), 1-minute or 2-minute-long; the continuous taking of short counts allowed us to aggregate these short counts into longer counts, as needed for the study.

5.2   Accuracy

The accuracy of GammaSense was evaluated by emulation on the collected data set. The technique of emulation tests the machine-learning algorithms in an environment emulated by the data set, for the ability to distinguish among indoor horizontal locations, indoor vertical locations and indoor versus outdoor. The emulations were run in the Weka tool (as described in Section 4) using fivefold cross-validation for splitting up the data sets into training and test data for the machine-learning algorithms. The features used in the emulations were 60-second gamma counts, and the mean and variance of the last five 60-second counts.

The performance of the localization algorithm is judged in the rest of this section based on quantitative measures, i.e. the *detection accuracy*, the *confusion matrix* and the *kappa statistic*. The detection accuracy is the percentage of correctly classified tests. A confusion matrix shows how many instances have been assigned to each class; the matrix elements show the percentage of test examples whose actual class (i.e. the actual location) is the row and whose predicted class (i.e. the predicted location) is the column.

The kappa statistic quantifies how much a detector is an improvement over a random detector. It can be thought of as the chance-corrected prediction agreement, and possible values range from +1 (perfect agreement between prediction and reality) via 0 (no agreement above that expected by chance) to -1 (complete disagreement).

Horizontal. The aim for the indoor horizontal localization was to distinguish among different building segments based on their different gamma patterns. The ComLab building consists of three wings built by 1901, 1993 and 2006, respectively, with the oldest two wings highly similar in style and material. We grouped our ComLab data into three classes, one for each building part. The emulation for this three-class recognition problem gave a detection accuracy of 67.7% and a kappa statistic of 0.51.

Further analysis of the prediction errors revealed that they were not distributed evenly between the three classes, as shown by the confusion matrix in Table 4(b). The highest confusion rates were between the 1901 and 1993 wings, a fact we explain by a high similarity in building materials and style, despite the different ages. Furthermore, if the data is regrouped into a two-class problem for distinguishing between the 1901/1993 and 2006 wings, the accuracy increases to 81.7% and the kappa statistic to 0.60, as summarized in Table 4(a). Hence, localization accuracy depends on similarity of construction. We conclude that GammaSense distinguishes between building segments of different construction parameters with moderate performance.

Table 4. ComLab horizontal results

(a) Detection results for localization among the three wings (first row). The same if the oldest two wings are aggregated into a single wing (second row).

	Accuracy	Kappa
ComLab (3 wings)	67.7%	0.51
ComLab (2 wings)	81.7%	0.60

(b) Confusion matrix for the 3-wing study. Elements show the percentage of tests whose row is the actual wing and whose column is the predicted wing.

	1901	1993	2006
1901	15.6%	9.7%	1.1%
1993	8.1%	23.1%	7.5%
2006	1.1%	4.8%	29.0%

Vertical. For the indoor vertical localization, the target was floor detection. The Egaa, ComLab, Math, and CS data sets were used in this evaluation, and the emulation for these buildings gave the results in Table 5(a). Overall, the Egaa domestic house allowed us a high degree of floor prediction agreement, while this decreased for the three public institutions.

From the confusion matrix for the vertical ComLab study (Table 5(b)) one identifies that one particular reason for the poor accuracy is the fact that the fourth and basement floors are comparable in gamma counts. While high counts were expected for the basement floor, one explanation for the high counts of the top floor is the fact that either the roof of the building is highly radioactive, or that the poor building ventilation has the radioactive gases accumulate under the roof. The Math building does not suffer from this issue and exhibits higher accuracy and kappa statistic. CS gave the poorest results for the kappa statistic, possibly because the CS building is new, the shielding between floors and from the subjacent earth is intact, and there exists an active ventilation system.

We then conclude that GammaSense distinguishes among floors in a building, yet there exists a set of parameters which decrease detection accuracy, such as roof accumulation of radon or good ventilation.

Indoor versus Outdoor. An additional aim for GammaSense was to detect indoor versus outdoor locations. Indoor and outdoor data from ComLab was used in the evaluation; the outdoor data was collected at two locations: one in

Table 5. Indoor vertical results

(a) Results for each testbed quantified by detection accuracy and kappa statistic.

	Accuracy	Kappa
Egaa	72.2%	0.45
ComLab	43.4%	0.29
Math	48.6%	0.35
CS	49.2%	0.23

(b) Confusion matrix for ComLab vertical. Elements show the percentage of tests whose row is the actual floor and column the predicted floor.

	0	2	3	4
0	10.5%	1.3%	0.0%	9.2%
2	0.0%	5.3%	6.6%	9.2%
3	0.0%	5.3%	9.2%	6.6%
4	9.2%	0.0%	0.0%	11.8%

the atrium (a small open yard in the core of the building), and the other on the lawn in front. The emulation results were:

$$\text{accuracy: } 91.7\% \text{ and kappa statistic: } 0.55.$$

We hypothesize that the better "ventilation" outdoor, either in the absence of building materials (which is the case for the lawn location) or even in their presence (the case for the inner atrium, surrounded by walls and paved) resulted in lower counts outdoors.

Sensitivity Analysis As a further analysis, we looked into more detail over the localization accuracy of the GammaSense algorithm, in the ComLab horizontal study for detecting among the three building segments, which gave the results on the first row of Table 4(a).

Specifically, we tested the variation of the detection accuracy and the kappa statistic as a function of the window size (i.e. the number of the last 60-second measurements whose mean and variance are given, together with the current reading, to the localization algorithm; the value for all the results reported above was five). The variation is reported in Fig. 5. As expected, since radiation counts are fairly unstable in value from one count to the other (as clear in Fig. 1 and 2), localization accuracy improves by considering more measurements per location; up to a number of eight minute-counts, performance parameters increase.

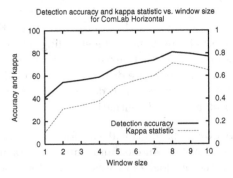

Fig. 5. The variation of detection accuracy and kappa statistic with window size (i.e. number of 1-minute measurements taken at each location)

6   Conclusions and Future Work

We investigated the performance of a machine-learning algorithm to pinpoint roughly the interfloor, intrafloor and indoor-versus-outdoor position of a client sensing background radioactivity levels indoors. We found that the accuracy results vary widely with the structure of the fingerprinted building, and report best performance for interfloor detection in a domestic house (a 72.2% detection accuracy) and intrafloor detection in a complex building made up of segments with different construction parameters (a 67.7% accuracy in wing detection). Also, we verify that the indoor-versus-outdoor detection for outdoor locations in the vicinity of the building has good performance (91.7%). Finally, we look into the variation of performance parameters with the window size (i.e. the number of minutes a client has to measure radiation levels at a location to have it detected) and report that performance increases with increasing window size, up to 8 minute-counts.

While we argue that harvesting natural parameters such as background radiation for location detection is worthy of research due to ease of deployment and a complete independence of infrastructure, we recognize that such techniques offer less performance and control than more standard, infrastructure-based techniques for localization. We state that—due to the geometry of its sources indoors—background radioactivity is a desirable signal for use in location estimation, especially one based exclusively on natural signals such as natural light, sound or the chemical components in the air (a potential future work).

The Geiger-Müller tube and counter that we used in our experiments are bulky for real-world deployments; however, miniaturized versions already exist—for instance, such sensors were integrated into mobile phones for early detection and localization of radioactive threats [16].

References

1. Kjærgaard, M.B.: A Taxonomy for Radio Location Fingerprinting. In: Proceedings of the Third International Symposium on Location- and Context-Awareness (2007)
2. Bahl, P., Padmanabhan, V.N.: RADAR: An In-Building RF-based User Location and Tracking System. In: Proceedings of the 19th Annual Joint Conference of the IEEE Computer and Communications Societies, INFOCOM (2000)
3. Otsason, V., Varshavsky, A., Marca, A.L., de Lara, E.: Accurate GSM Indoor Localization. In: Beigl, M., Intille, S., Rekimoto, J., Tokuda, H. (eds.) UbiComp 2005. LNCS, vol. 3660, pp. 141–158. Springer, Heidelberg (2005)
4. Youssef, M., Agrawala, A.: The Horus WLAN Location Determination System. In: Proceedings of the Third International Conference on Mobile Systems, Applications, and Services (2005)
5. Haeberlen, A., Flannery, E., Ladd, A.M., Rudys, A., Wallach, D.S., Kavraki, L.E.: Practical Robust Localization over Large-Scale 802.11 Wireless Networks. In: Proceedings of the Tenth ACM International Conference on Mobile Computing and Networking (2004)

6. Varshavsky, A., Lamarca, A., Hightower, J., de Lara, E.: The SkyLoc Floor Localization System. In: Proceedings of the Fifth Annual IEEE International Conference on Pervasive Computing and Communications (2007)

7. Patel, S., Truong, K., Abowd, G.: PowerLine Positioning: A Practical Sub-Room-Level Indoor Location System for Domestic Use. In: Dourish, P., Friday, A. (eds.) UbiComp 2006. LNCS, vol. 4206, pp. 441–458. Springer, Heidelberg (2006)

8. Ravi, N., Iftode, L.: FiatLux: Fingerprinting Rooms Using Light Intensity. In: Adjunct Proceedings of the Fifth International Conference on Pervasive Computing (2007)

9. Draganić, I.G., Draganić, Z.D., Adloff, J.-P.: Radiation and Radioactivity on Earth and Beyond, 2nd edn. CRC Press, Boca Raton (1993)

10. Royal Society of Chemistry: Essays on Radiochemistry: Alpha, Beta and Gamma Radioactivity (unknown year),
    http://www.rsc.org/pdf/radioactivity/number3.pdf

11. United Nations Scientific Committee on the Effects of Atomic Radiation (UNSCEAR): ANNEX B: Exposures from natural radiation sources, subsubsection IIC2 (2000), http://www.unscear.org/docs/reports/annexb.pdf

12. European Commission: Commission recommendation of 21 february 1990 on the protection of the public against indoor exposure to radon (1990),
    http://eur-lex.europa.eu/LexUriServ/LexUriServ.do?uri=
    CELEX:31990H0143:EN:NOT

13. Ghany, H.A.A.: Variability of radon levels in different rooms of egyptian dwellings. Indoor and Built Environment 15(2), 193–196 (2006)

14. Sonkawade, R.G., Ram, R., Kanjilal, D.K., Ramola, R.C.: Radon in tube-well drinking water and indoor air. Indoor and Built Environment 13(5), 383–385 (2004)

15. Witten, I.H., Frank, E.: Data Mining: Practical machine learning tools and techniques, 2nd edn. Morgan Kaufmann, San Francisco (2005)

16. Venere, E., Gardner, E.K.: Cell phone sensors detect radiation to thwart nuclear terrorism (2008),
    http://www.purdue.edu/UNS/x/2008a/080122FischbachNuclear.html

People Identification Using Gait Via Floor Pressure Sensing and Analysis

Gang Qian[1,2], Jiqing Zhang[1,2], and Assegid Kidané[1]

[1] Arts, Media and Engineering Program and
[2] Department of Electrical Engineering
Arizona State University, Tempe, AZ 85287, USA
{Gang.Qian,Jiqing.Zhang,Assegid.Kidane}@asu.edu

Abstract. This paper presents an approach to people identification using gait based on floor pressure data. By using a large area high resolution pressure sensing floor, we were able to obtain 3D trajectories of the center of foot pressures over a footstep which contain both the 1D pressure profile and 2D position trajectories of the COP. Based on the 3D COP trajectories a set of features are then extracted and used for people identification together with other features such as stride length and cadence. The Fisher linear discriminant is used as the classifier. Encouraging results have been obtained using the proposed method with an average recognition rate of 94% and false alarm rate of 3% using pairwise footstep data from 10 subjects.

Keywords: Pressure analysis; gait recognition; biometrics.

1 Introduction

Recognizing people using gait is an important area in biometrics with applications in homeland security, access control, and human computer interaction. While most of research for people identification using gait has been focused on computer vision based techniques, there has been research addressing gait recognition using foot pressure information, for example [1, 4, 5, 6, 8, 9, 10, 11, 12]. Orr and Abowd [8] have researched on people identification based on the pressure profile over time during a foot step on a load-cell sensor. The footstep profile features used in these approaches include the mean, the standard deviation, and the duration of the pressure profile, the overall area under the profile, and pressure value and the corresponding time of some key points such as the maximum point in the first and last halves of the profile and the minimum point between them. In these approaches, due to the nature of load-cell based pressure sensing, the spatial pressure distribution during a foot step is not measured and only the amount of pressure and the corresponding time can be used for feature extraction and people identification. Therefore, the load-cell based gait identification approaches do not take into account the spatial pressure distribution and features such as the trajectories of center of pressure (COP). On the other hand, Jung et al. [5, 6] used a mat-type pressure sensor for gait recognition based only on the 2D COP trajectory. To recognize a person based on the foot pressure measured during a gait cycle, it is important to extract both spatial features such as the trajectories of COP and footstep pressure profile features. To address this issue, in this paper we present a gait-based

D. Roggen et al. (Eds.): EuroSSC 2008, LNCS 5279, pp. 83–98, 2008.

people identification method using foot pressure obtained using a pressure sensing floor. Our goal is to develop a system that can reliably identify a subject group of 5 to 20 people. The pressure sensing floor system [14, 15] we use consists of a number of pressure sensing mats from Tekscan arranged in a rectangular shape spanning a total sensing area of about 180 square feet. Each pressure sensing mat has 2016 force sensing resistor (FSR) based sensors in a resolution of over 6 sensors per square inch. By using this large area, high resolution pressure sensing floor, we extract features both from the trajectories of the COP and the pressure profile of both left and right foot steps during a gait cycle. We also use other gait features such as the stride length and cadence as features. In our approach, the Fisher linear discriminant is used as the classifier. Encouraging results have been obtained using the proposed method with an average recognition rate of 94% and false alarm rate of 3% using pair-wise footstep data from 10 subjects. Experimental results show that the proposed method achieves better or comparable performance compared to existing methods.

However, the proposed approach is still limited in a few aspects. In our research on people identification using floor pressure, we only consider the case that people walking with their shoes off. We also assume that people are walking in a straight line at a normal speed. In addition to these limitations, the proposed approach still needs to be further evaluated, e.g. by using floor pressure data collected with various walking speeds, and a large dataset with more subjects.

2 Preprocessing of Floor Pressure Data

To recognize people using gait based on floor pressure, it is necessary to first pre-process the raw pressure data obtained using the pressure sensing floor and then extract features that can be used for gait recognition. A number of tasks need to be accomplished by this preprocessing step, including pressure data clustering, tracking of the cluster centers, recognition of left and right feet, estimation of walking directions, and rectification of the COP trajectories.

2.1 Clustering and Tracking of Centers of Pressure Using Mean-Shift

The mean-shift algorithm [2] is used to cluster floor pressure data and track the center of pressure over time. Mean-shift is a repetitively shifting process to find the sample mean of a set of data samples. In our case the data samples are the 2D locations of points on the floor with active pressure readings. The mean-shift vector $M_h(x)$ at x using a kernel $G(x)$ can be found using the following equation:

$$M_h(x) \equiv \frac{\sum_{i=1}^{n} G(\frac{x_i - x}{h}) w(x_i)(x_i - x)}{\sum_{i=1}^{n} G(\frac{x_i - x}{h}) w(x_i)} \tag{1}$$

where $w(x_i)$'s are the sample weights and h is the window size for the kernel. In our experiments, the positions of pressure data points are firstly clustered through the blurring process [2] using a Gaussian kernel. The observed pressure at a point is then used as the corresponding weight for this point. Once the process has converged, the

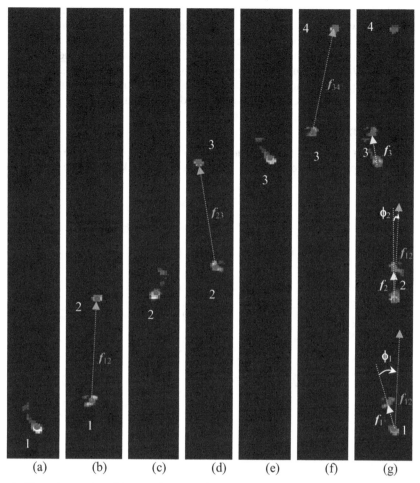

    (a)   (b)   (c)   (d)   (e)   (f)   (g)

Fig. 1. Clustering and tracking results over about 1.5 gait cycles. (a) through (f) show snap shots of observed floor data and corresponding COPs (red dots) and their ID numbers (white digits). It can be seen that the IDs of the COP remain unchanged during a single footstep. The inter-foot vectors are shown by the green arrows. (g) shows overlapped floor data from (a) to (f) so that the intra-foot vectors can be visualized (the yellow arrows) properly.

data set will be tightly packed into clusters, with all of the data points located closely to the center of that cluster. The process is said to have converged either after the maximum number of iterations defined by the algorithm is reached or earlier when the mean shift of centers becomes less than the convergence threshold. After convergence, each cluster is assigned with a unique cluster ID number and every data point has a 'label' associated with corresponding cluster. For every subsequent pressure data frame, centers from the previous frame are updated through the mean shift algorithm (1) using current observed pressure values as weights and checked for convergence. In practice, entirely new data points resulting in new cluster centers can occur if there are groups of data points not assigned to any existing cluster centers. Figure 1 shows clustering and tracking results of the pressure data over about 1.5 gait cycles.

The red dots indicates the cluster centers and the white digits their ID numbers. During a *footstep*, which is defined as the period between the heel-strike and the toe-off of a foot, the COP of the foot can be correctly tracked using the mean-shift algorithm since the corresponding cluster ID is maintained. In addition, we define the *pressure related to a COP* as the sum of the pressure values of all the samples in the corresponding cluster. As a result, over a footstep, a 3D COP trajectory can be obtained, including 2D position trajectories of the COP and the pressure profile over time.

2.2 Foot Recognition

It is important to separate the left and right feet so that proper features can be extracted for people identifications. In our research, we take a simple and efficient

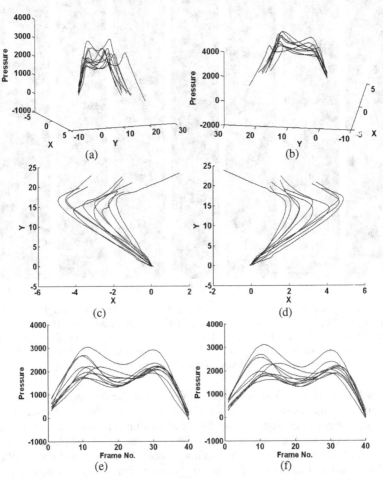

Fig. 2. Mean COP trajectories of the left (left panels) and right (right panels) footsteps using data from 10 subjects. (a) and (b) are the 3D COP trajectories including pressure and position information. (c) and (d) are the rectified 2D position COP trajectories and (e) and (f) the 1D pressure profile.

approach to foot recognition based on the COP trajectories obtained using the mean-shift algorithm. Let $C_1=\{c_1(s_1),...,c_1(e_1)\}$ and $C_2=\{c_2(s_2),...,c_2(e_2)\}$ be two successively detected and tracked COP trajectories over two adjacent footsteps, footstep 1 and footstep 2, where $c_1(s_1)$ and $c_1(e_1)$ represent the position of the heel-strike and toe-off of one foot, and $c_2(s_2)$ and $c_2(e_2)$ that of the other foot, and s and e are time indices. The goal is to identify which footstep corresponds to the left foot and which one to the right foot. To do that, we define two types of vectors, namely the intra-foot vectors and the inter-foot vectors. The intra-foot vectors are vectors from the heel-strike position to the toe-off position during a single footstep, for example, $f_1=(c_1(s_1)\rightarrow c_1(e_1))$ and $f_2=(c_2(s_2)\rightarrow c_2(e_2))$. The inter-foot vector is from the toe-off of the first footstep to the heel-strike of the second footstep. In this example, $f_{12}=(c_1(e_1)\rightarrow c_2(s_2))$. Recall that we assume people walk in straight line.

Let $\phi_1=\angle f_1 f_{12}$, and $\phi_2=\angle f_2 f_{12}$ be the clockwise acute angles (possibly negative) from the intra-foot vectors to the inter-foot vector. By using ϕ_1 and ϕ_2 it is easy to separate the left foot from the right foot since when footstep 1 is the left foot, $\phi_1>\phi_2$ and vice versa. See Figure 1 for an example.

2.3 Rectification of COP Trajectories

For each foot step, a 3D COP curve can be obtained as a function over time, which consists of the 2D position trajectories and the corresponding floor pressure trajectories. After the separation of the left and right feet, the 3D COP curves of both feet spanning over a gait cycle can be obtained. At this point, the 2D position trajectories are in a global floor-centered coordinate frame. To make the COP trajectories invariant to the walking direction of the person, a rectification step is needed to rotate the COP trajectories so that the resulting trajectories are represented in a local foot-centered coordinate frame. In our research, we first estimate the walking direction as

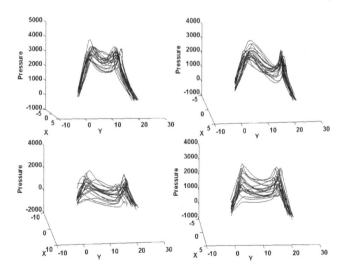

Fig. 3. Samples of the 3D COP trajectories of the left (left panels) and right footsteps from the seventh (first row) and the tenth subjects

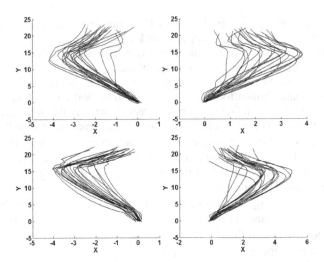

Fig. 4. Samples of the 2D COP trajectories of the left (left panels) and right footsteps from the seventh (first row) and the tenth subjects

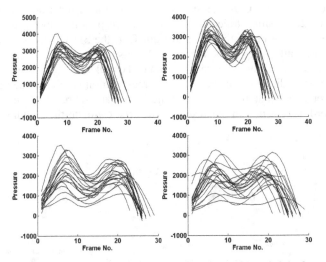

Fig. 5. Samples of the 1D pressure profiles of the left (left panels) and right footsteps from the seventh (first row) and the tenth subjects

the mean of the left and right intra-foot vectors over a single gait cycle. This walking direction is then used as the y-axis direction of the local coordinate frame and the x-axis direction is easily determined by the right-hand-rule so that the resulting z-axis (also the pressure-axis) is up. The origin of the local foot-centered frame is taken as the heel-strike point of the corresponding footstep. Figure 2 shows the mean plots of the COP and pressure trajectories of the left (a) and right (b) footsteps of 10 subjects. Panels (c) (d) and (e) (f) are the projected 2D position and 1D pressure trajectories. Figures 3 to 5 respectively show samples of the complete 3D COP and 2D position COP trajectories, and 1D pressure profiles of the left (left panels) and right footsteps

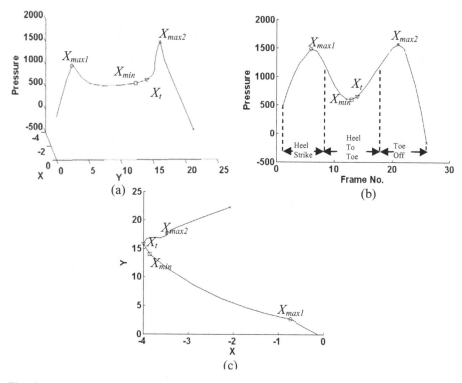

Fig. 6. A typical 3D COP trajectory of a left footstep (a), and the related 1D pressure (b) and 2D position (c) trajectories. The corresponding key points are also labeled.

from the seventh (first rows) and the tenth subjects. From these plots, it can be seen that there is certain inconsistency in the COP trajectories of the same subject from trial to trial. Meanwhile, similarity exists in the COP trajectories of the two subjects. Thus it is important to extract reliable features from the 3D COP trajectories.

3 Feature Extraction

It is obvious that the observations we have, i.e. the 3D position-pressure COP trajectories contain all the information used by existing methods, including the footstep-based approaches [8], and the 2D position COP based approaches [5, 6]. It is expected that by using the comprehensive 3D COP trajectories improved people identification results can be achieved. People identification using gait based on floor pressure is essentially a pattern classification problem, where the dynamic foot pressure distribution of a walking subject is a pattern that need to be recognized and then the subject can be identified. Such a pressure distribution pattern is reflected by the two 3D COP curves during the left and right footsteps. To tackle this classification problem, we need to extract features from the 3D COP trajectories. In our research, we first identify a set of key points along a 3D COP trajectory according to the pressure and

Table 1. Definition of key points of a footstep

Key point	Definition
Xmax1	The first dominant local maxima in the pressure trajectory
Xmax2	The second dominant local maxima in the pressure trajectory
Xmin	The local minima in the pressure trajectory between Xmax1 and Xmax2
Xt	The turning point in the 2D position COP trajectory
Xend	The end of the entire COP trajectory

position values of the COP. These key points are defined in Table 1. The key points of a typical COP trajectory of a left footstep are shown in Figure 6. Among these key points, Xmax1, Xmax2 and Xmin are also used by the footstep-based methods, e.g. [7].

Within a footstep, Xmax1 corresponds to the toe-off moment of the other foot. Xmin appears during full foot contact on the floor. Xmax2 takes place when the heel of the other foot is about to strike on the floor. Xt is defined when the absolute value of the x coordinate (orthogonal to walking direction) of COP reach maximum. Xt reflects the degree of wiggle of the subject during walking. For each key point, the corresponding position, pressure value, and curve length (measured from the start of the COP trajectory), relative frame numbers (counted from the start of the COP) can be used as features. Such key-point based features from a pair of successive (e.g. right and left) footsteps are used as a feature vector for people identification. In our experiments, we tried different feature combinations and compared their performances. In addition to these features, the sum of the mean pressure of such a pair of successive footsteps is also used as a feature that encodes the body weight of the subjects. Furthermore, stride length and stride cadence are also extracted as part of the feature set.

4   People Identification Using Fisher Linear Discriminant

Many machine learning algorithms exist that can solve supervised pattern classification problems, including the Bayesian classifiers, the k-nearest neighbors (KNN) method, the support vector machine (SVM) method, and the linear discriminant analysis (LDA) method, just to name a few. An overview of these methods can be found in a standard machine learning textbook, e.g. [3]. In our experiments, we selected the binary version of the LDA, namely, the Fisher linear discriminant (FLD) method as the classifier, due to its simplicity in training and testing. We made use of a Matlab statistic pattern recognition toolbox [13] for FLD training and testing.

An M-class classification problem can be solved by training M FLD classifiers, one for each class in a one-vs.-the-rest framework. For each FLD classifier, training data are split into an in-class sample set which has samples in the corresponding class and an out-of-class sample set with the remaining training samples of other classes. In the training of an FLD classifier, an optimal linear transformation parameterized by (W,b) needs to be found to maximize the between-class separability and to minimize the within-class variability, thus achieving maximum class discrimination. The parameters (W,b) can be found by solving a generalized eigenvalue problem. Once (W,b) are found, whether a test point x is in the class or not is based on the sign of the discriminant function $q(x)$ given by

$$q(x) = W^T x + b \tag{2}$$

When $q(x)$ is nonnegative, the test point is classified as an in-class data, otherwise, it is not in the class.

5   Experimental Results and Performance Analysis

5.1   Data Collection

In our experiments, we collected floor pressure data from 10 subjects, 8 men and 2 women, with an age range of 24 to 50, and the height range 165cm to 180cm. For each subject, we collected about 2-4 minutes of pressure data during normal walking. After foot recognition and COP tracking using mean-shift, 36 to 76 pairs of 3D COP trajectories are extracted from pairs of successive footsteps for each subject. In total, 529 pairs of 3D left and right foot COPs were used in training and testing.

5.2   Feature Selection

In Section 3, a number of features are introduced for people identification. Some of the features are related to the key points such as the locations, pressure values, curve lengths, and frame number of the key points, and some of them contain summarizing features of the subject's gait, such as the stride length and stride cadence. In our experiments, we tried different feature sets and compared their performances. There are a number of things we have in mind when we chose the features. The first thing is whether the feature set is invariant to walking speed. It is ideal to have a people ID system that can recognize the subject across changing walking speed. Although the current data set was collected at normal walking speeds of individual subjects, it is still valid to compare the performance of a speed-invariant feature set and that of a speed-variant feature set. The second thing we want to see is how important the pressure values are in people identification. As a result, we tested the performance of the following four sets of features.

Feature sets 1 makes use of the relative frame numbers of the key points and the stride cadence, which are tightly related to the walking speed. Therefore, FS-1 is considered to be a speed-variant feature set. FS-2 uses normalized frame numbers of the key points by the frame length of the corresponding gait cycles. Thus it is less dependent on the walking speed. However, since FS-2 includes the stride length, it is considered to be partly speed-variant. On the other hand, FS-3 removes the stride length as a feature, it is thus speed-invariant. Finally, all the pressure values are removed from FS-3 to get FS-4, which is both speed-invariant and free from using pressure values.

5.3   Cross Validation and Performance Analysis

To validate the performance of different feature sets and obtain an overall picture of the proposed approach, we ran cross-validation over the entire data set. One hundred runs were taken. For each run, a random sample set of 20 COP pairs was drawn from the data set of each subject. The remaining samples were then used as testing data set. In each run, using the same training and testing sets, the results obtained using the

Table 2. Four feature sets

Feature set ID	Key points included	Features used for each key point	Other features	Invariant to speed?	Pressure values used?
FS-1	Xmax1, Xmax2, Xmin, Xend	- rectified position, - pressure, - curve length, - relative frame no.	-stride length -stride cadence - mean pressure of both footsteps	No	Yes
FS-2	Xmax1, Xmax2, Xmin, Xend	- rectified position, - pressure, - curve length, - relative frame no. normalized by the number of frames of related gait cycle,	-stride length - mean pressure of both footsteps	somewhat	Yes
\FS-3	Xmax1, Xmax2, Xmin, Xend	- rectified position, - curve length, - pressure - relative frame no. normalized by the number of frames of related gait cycle,	- mean pressure of both footsteps	Yes	Yes
FS-4	Xmax1, Xmax2, Xmin, Xend	- rectified position, - curve length, - relative frame no. normalized by the number of frames of the related gait cycle,		Yes	No

four feature sets were compared. The recognition rate (RR) and the false alarm rate (FAR) are used to evaluate the performance of the proposed people identification method. The RR is defined as the ratio of the number of correct recognitions (NCR) and number of total in-class data (NTI). The FAR is the ratio between the number of false alarms (NFA) and the number of total out-class data (NTO).

$$RR = \frac{NCR}{NTI} \times 100\% \ , \ FAR = \frac{NFA}{NTO} \times 100\% \tag{3}$$

Figure 7 shows the resulting RR (top panel) and FAR over 100 trials using the four feature sets. The results for each trial are the average RR and FAR of the 10 subjects. It can be seen that FS-1 and FS-2 have the highest RR and lowest FAR, which is reasonable since both of them includes most of the features. This result indicates that using the normalized timestamp (i.e. frame number) of the key points has no noticeable effect on the performance. The performance of FS-3 drops compared to those of FS-1 and FS-2, which means that the stride length is a contributing feature to recognition. In addition, it is clear from Figure 7 that the FS-4 has the worst performance indicating that it is important to include pressure values as part of the feature vector.

The average performances of the proposed methods with the four feature sets for each person are presented in Tables 3 and 4. It can be seen that FS-1 and FS-2 share the same RR while FS-2 has slightly larger FAR than that of the FS-1. The performance of FS-4 is the worst among the four feature sets. While the performance of FS-3 is in the middle of the performances of feature sets 1 and 2, and FS-4.

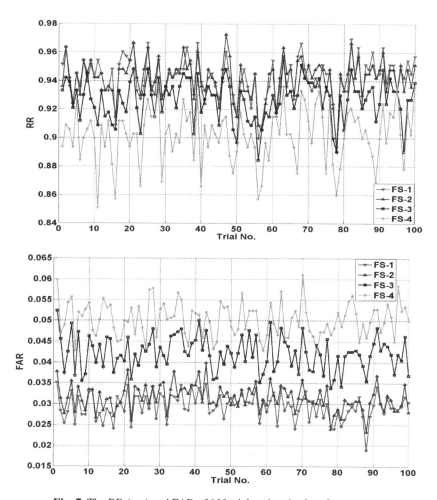

Fig. 7. The RR (top) and FAR of 100 trials using the four feature sets

In addition to the comparison among the four feature sets, we also compared the results of the proposed method to those of the existing people identification methods using floor pressure. The comparison is summarized in Table 5. It can be seen that according to the features used, existing methods can be roughly clustered into methods using only pressure profiles [1, 8, 9, 10], methods using only 2D positional COP trajectories [5, 6], and methods based on explicit gait features such as stride length, cadence [7, 11, 12]. In general, the method only using 2D COP performed the worst among all the methods. Excellent people recognition results have been reported using the pressure profile [8] and gait features [11, 12]. The features used in the proposed method explore the 3D COP trajectories which contains both the pressure profiles and the 2D positional trajectories. In addition, stride length and cadence are also used in the proposed method. Therefore, the proposed method makes use of a nearly comprehensive feature sets which include nearly all the key features of the existing methods. This is made possible by using a high-resolution pressure sensing floor and robust

Table 3. Average recognition results of FS-1 and FS-2 over 100 trials

Subject ID#	NTI	Recognition				NTO	False Alarm			
		FS-1		FS-2			FS-1		FS-2	
		NCR	RR %	NCR	RR %		NFA	FAR %	NFA	FAR %
1	26	25.9	99.6	25.9	99.5	303	3.3	1.09	3.21	1.06
2	51	46.7	91.6	46.7	91.6	278	13.5	4.87	17.1	6.15
3	56	55.3	98.7	55.4	98.8	273	7.72	2.83	7.12	2.61
4	16	15.4	96	15.3	95.8	313	5.47	1.75	5.51	1.76
5	34	31.4	91.6	30.8	90.6	295	20.8	7.04	21.8	7.4
6	44	37.7	85.7	38	86.3	285	7.37	2.59	7.41	2.6
7	22	22	100	22	100	307	1.32	0.43	2.2	0.7
8	35	33.5	95.7	33.5	95.7	294	6.54	2.22	8.19	2.79
9	25	24.7	98.8	24.7	98.7	304	5.2	1.71	5.07	1.67
10	20	17.7	88.5	17.8	88.8	309	15.9	5.13	15.7	5.07
Overall	**329**	**310**	**94.2**	**310**	**94.2**	**2961**	**87.1**	**2.94**	**93.3**	**3.15**

Table 4. Average recognition results of FS-3 and FS-4 over 100 trials

Subject ID#	NTI	Recognition				NTO	False Alarm			
		FS-3		FS-4			FS-3		FS-4	
		NCR	RR %	NCR	RR %		NFA	FAR %	NFA	FAR %
1	26	25.9	99.6	25.7	99.5	303	4.1	1.35	5.8	1.91
2	51	44.6	87.5	40.9	91.6	278	22.4	8.04	33.3	12
3	56	55.3	98.8	55.2	98.8	273	9.61	3.52	10.3	3.77
4	16	15.3	95.8	14.7	95.8	313	5.49	1.75	4.97	1.59
5	34	30.3	89	29.1	90.6	295	26.8	9.08	31.3	10.6
6	44	36.7	83.5	35.4	86.3	285	9.26	3.25	16.3	5.7
7	22	21.9	99.8	21.8	100	307	12.9	4.21	12.5	4.06
8	35	33.4	95.5	33	95.7	294	8.34	2.84	9.78	3.33
9	25	24.8	99	24.3	98.7	304	4.94	1.63	4.72	1.55
10	20	16.6	83.2	16.7	88.8	309	21	6.8	20.2	6.52
Overall	**329**	**305**	**92.7**	**297**	**90.2**	**2961**	**125**	**4.21**	**149**	**5.03**

pressure clustering and tracking algorithm using mean-shift. At a result, the performance of the proposed approach is superior to nearly all the existing methods. Furthermore, the proposed method achieved top performances compared to the existing methods. None of the results of the existing methods was based on cross-validation as what we have done in our research. Hence it is not clear whether the reported recognition results of the existing methods were based on results of a number of trials or just one trial. The performance of an algorithm can vary by different selection of training and testing data set. In the second column to the right of Table 5, we have included the ranges of the RRs and FARs of our proposed methods. It can be seen that the peak RRs for FS-1 and FS-2 are equal or above 97%.

Table 5. Performance comparison of methods for people identification using floor pressure

Method	Floor sensor	Major features	Classi-fier	# of sub.	RR (%)	FAR (%)	Cross Validated ?	Invariant to speed?
Addlesee et al., 1997, [1]	Load-cells	Pressure profile over a footstep	HMM	15	<50	N/A	No	Possible
Jung, et al., 2003, [5]	Pressure mats	2D trajectories of COP	HMM	8	64	5.8	No	Possible
Pirttikangas et al., 2003, [9]	Pressure floor	Pressure profile over the entire floor during walking	HMM	3	76.9	11.6	No	Possible
Pirttikangas et al., 2003, [10]	Same as [9]	Same as the above [9]	Learn-ing VQ	11	<78	N/A	No	Possible
Jung, et al., 2004, [6]	Pressure mats	2D positional trajectories of COP	HMM-NN	11	79.6	2.05	No	Possible
Middleton et al., 2005, [7]	Force sensitive resistor (FSR) mats	Stride length, stride cadence, heel-to-toe ratio	N/A	15	80	N/A	No	No
Yun et al., 2003, [12]	Floor w/ on/off switch sensors,	Compensated foot centers over 5 consecutive footsteps	NN	10	92.8	N/A	No	No
Orr et al., 2000, [8]	Load-cells	Key points from pressure profile	KNN	15	93	N/A	No	No
Yoon et al., 2005, [11]	Floor w/ on/off switch sensors,	Compensated foot centers and heel-strike and toe-off time over 5 consecutive footsteps	NN	10	96.2	N/A	No	No
Proposed-FS-1	**FSR mats**	**See Table 2**	**FLD**	**10**	**94.2**	**2.94**	**Yes** RR: 89.4-97 FAR:1.89-3.5	**No**
Proposed-FS-2	**FSR mats**	**See Table 2**	**FLD**	**10**	**94.2**	**3.15**	**Yes** RR:89.1-97.2 FAR:2.33-4	**Possible**
Proposed-FS-3	**FSR mats**	**See Table 2**	**FLD**	**10**	**92.7**	**4.21**	**Yes** RR:88.5-95.7 FAR:3.38-5.2	**Yes**
Proposed-FS-4	**FSR mats**	**See Table 2**	**FLD**	**10**	**90.2**	**5.03**	**Yes** RR:85.1-93.6 FAR:4.26-6.1	**Yes**

Yoon et al. reported (non cross-validated) results [11] that seems to be better than ours (average after cross-validation). A close look reveals that this might not be the case. The system in [11] was restricted in the sense that a subject has to walk along a floor strip in a fixed direction at a normal speed. Furthermore the subject needs to walk on the floor for at least five steps. Features over five steps are concatenated to form a large feature vector for recognition. In contrast, our proposed method can recognize people walking in any direction. Once a pair of successive footsteps is captured, recognition can be conducted. If our approach were given data from five consecutive footsteps to recognize a person, the recognition rate can be much higher. Five pairs of left and right COP trajectories can be extracted from five steps. Assume that a person is recognized only when three or more pairs are correctly recognized. Let $P=0.927$ be the current recognition rate (FS-3) using a single COP pair of the proposed approach. If we treat the five pairs of data to be independent, the probability that the subject can be correctly recognized using five pairs of COPs can be computed as $P5=C_5^3 P^3(1-P)^2 + C_5^4 P^4(1-P) + C_5^5 P^5 = 0.9965$. Similarly the FAR can be computed in this case as 0.06%. However, the independency of the five left-right COP trajectory pairs is not true. Thus, in practice, the recognition rate of a subject using data from five steps can be lower than P5. On the other hand, intuitively using more data will improve the performance. Thus we expect the recognition rate of our approach using data from five steps will be somewhere between P and P5. More experiments need to be carried out to further compare the recognition performance of our approach and that reported in [11].

In addition, our proposed method takes into considerations of the speed-invariance, which can possibly recognize people at different walking speed. Another advantage of our proposed method is that it uses Fisher linear discriminant, which is a much more computationally efficient method than those used by existing methods such as hidden Markov models and neural networks. This makes the training and testing of the proposed method very fast. Using Matlab, it takes only 12.7 seconds to run the 100 trials with each trial including the training and testing of the four feature sets.

6   Conclusions and Future Work

A robust people identification approach using gait based on floor pressure is presented in this paper. Promising people identification results are obtained by using features extracted from the 3D COP trajectories in the pressure and position spaces. The proposed approach makes use of simple linear classifier for computational efficiency. This indicates that the features used in our approach from different people are mostly linearly separable. In the selection of the feature set, we also consider the issue of walking-speed invariance by normalizing the time stamps of the key points by the length of the corresponding gait cycle. Our experimental results show that by doing so the performance only slightly decreases. The proposed approach achieves better or comparable people recognition results compared to existing methods for people identification using floor pressure.

One of the limitations of the reported results in this paper is that it is based on a relatively small data set from 10 subjects walking in straight line without shoes in a normal walking speed. As part of our future work, we will collect more data especially those with varying walking speeds/patterns from an enlarged testing population

(~20 people) to further test the performance of the proposed method in these scenarios. Furthermore, we will evaluate the system performance by using more foot steps to better compare the proposed approach to that in [11].

Pressure sensors have been integrated with other sensors including tile angle sensor, gyroscope, bend sensor, and accelerometer for people identification using gait [4]. Excellent people identification results have been reported in [4] based on a testing population of 9 subjects. The approach proposed in [4] makes use of pressure sensors of low spatial resolution. Another research direction we will pursue in the future is to integrate the high resolution pressure sensing floor with other sensors such as those used in [4] to further improve people identification using gait e.g. by increasing the size of testing population that the system can reliably recognize.

Acknowledgement

This paper is based upon work partly supported by U.S. National Science Foundation on CISE-RI no. 0403428 and IGERT no. 0504647. Any opinions, findings and conclusions or recommendations expressed in this material are those of the authors and do not necessarily reflect the views of the U.S. National Science Foundation (NSF).

References

1. Addlesee, M., Jones, A., Livesey, F., Samaria, F.: The ORL Active Floor. IEEE Personal Communications, 35–41 (October 1997)
2. Cheng, Y.: Mean Shift, Mode Seeking and Clustering. IEEE Transactions on Pattern Analysis and Machine Intelligence 17(8), 790–799 (1995)
3. Duda, R.O., Hart, P.E., Stork, D.G.: Pattern Classification, 2nd edn. John Wiley & Sons, New York (2000)
4. Huang, B., Chen, M., Huang, P., Xu, Y.: Gait Modeling for Human Identification. In: IEEE International Conference on Robotics and Automation, pp. 4833–4838 (2007)
5. Jung, J., Bien, Z., Lee, S., Sato, T.: Dynamic-Footprint Based Person Identification Using Mat-type Pressure Sensor. In: International Conference of the IEEE Engineering in Medicine and Biology Society, vol. 3, pp. 2937–2940 (2003)
6. Jung, J., Sato, T., Bien, Z.: Dynamic Footprint-Based Person Recognition Method Using a Hidden Markov Model and a Neural Network. International Journal of Intelligent Systems 19(11), 1127–1141 (2004)
7. Middleton, L., Buss, A.A., Bazin, A., Nixon, M.S.: A Floor Sensor System for Gait Recognition. In: IEEE Workshop on Automatic Identification Advanced Technologies, pp. 171–176 (2005)
8. Orr, R.J., Abowd, G.D.: The Smart Floor: A Mechanism For Natural User Identification and Tracking. In: Conference on Human Factors in Computing Systems, pp. 275–276 (2000)
9. Pirttikangas, S., Suutala, J., Riekki, J., Röning, J.: Footstep Identification From Pressure Signals Using Hidden Markov Models. In: Finnish Signal Processing Symposium, pp. 124–128 (2003)
10. Pirttikangas, S., Suutala, J., Riekki, J., Röning, J.: Learning Vector Quantization in Footstep Identification. In: IASTED International Conference on Artificial Intelligence and Applications, pp. 413–417 (2003)

11. Yoon, J., Ryu, J., Woo, W.: User Identification using User's Walking Pattern over the ubi-FloorII. In: International Conference on Computational Intelligence and Security, vol. 3801, pp. 949–956 (2005)
12. Yun, J., Lee, S., Woo, W., Ryu, J.: The User Identification System Using Walking Pattern Over the ubiFloor. In: International Conference on Control, Automation, and Systems, pp. 1046–1050 (2003)
13. Franc, V., Hlavac, V.: Statistical Pattern Recognition Toolbox for Matlab User's Guide. Research Report CTU-CMP-2004-08, Center for Machine Perception, K13133 FEE Czech Technical University (2004)
14. Rangarajan, S., Kidane, A., Qian, G., Rajko, S.: The Design of a Pressure Sensing Floor for Movement-Based Human Computer Interaction. In: Kortuem, G., Finney, J., Lea, R., Sundramoorthy, V. (eds.) EuroSSC 2007. LNCS, vol. 4793, pp. 46–61. Springer, Heidelberg (2007)
15. Rangarajan, S., Kidane, A., Qian, G., Rajko, S.: Design Optimization of Pressure Sensing Floor for Multimodal Human Computer Interaction. In: Advances in Human Computer Interaction, I-Tech Education and Publishing, Vienna, Austria (2008)

Location-Free Object Tracking on Graph Structures

Daniela Krüger, Carsten Buschmann, and Stefan Fischer

Institute of Telematics, University of Lübeck
Ratzeburger Allee 160, 23538 Lübeck, Germany
{krueger,buschmann,fischer}@itm.uni-luebeck.de
http://www.itm.uni-luebeck.de

Abstract. Using wireless sensor networks for object tracking requires ordering events with regard to time and location. In labyrinth-shaped topologies, one-dimensional ordering suffices within the different sections of the network. We present an algorithm that decomposes the network into such sections, tracks objects within using binary sensors and, if required, hands them over to the next section. We evaluate our approach through extensive simulations and show that it is robust against sensor failures and packet loss.

Keywords: In-building Object Tracking, Wireless Sensor Networks.

1 Introduction

Object tracking is a common application domain for wireless sensor networks. Most authors assume that the sensor nodes are spread out over a two dimensional, convex area. In such a setting, the typical approach of using localization algorithms to make all devices location-aware works relatively well. However, it was shown in [1] that such algorithms fail for settings where the network is non-convex, contains a large number of wholes, et cetera. A typical application where such non-convex networks occur is indoor object tracking. If the nodes are deployed in building corridors, the wireless sensor network consists of a topological structure resembling roads.

In this paper we consider the problem of tracking objects in such scenarios using binary sensors, i.e. sensors that can detect whether an object is present or not. We pick up the idea proposed in [1,2] to interpret the network structure as a graph. Corridor junctions and dead-ends are considered as graph vertices, whereas the corridor segments in-between (also called corridors) are considered as graph edges (cf. Figure 1). After network deployment, nodes determine whether they reside in a vertex or an edge area of the network. If they are part of an edge, i.e. reside in a corridor segment, they compute the one-dimensional ordering of their neighbors that is then used for object tracking.

Tracking is done on a 'per corridor' basis: Nodes within one corridor track the object until it reaches a junction, then tracking is handed over to nodes in the next corridor.

D. Roggen et al. (Eds.): EuroSSC 2008, LNCS 5279, pp. 99–111, 2008.
© Springer-Verlag Berlin Heidelberg 2008

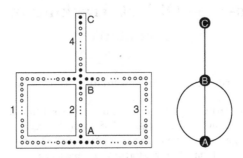

Fig. 1. Scenario

The remainder of this paper is structured as follows. In the next section we discuss related work. In Section 3 we describe our approach for one-dimensional object tracking comprising start-up and operating phase. The two phases are evaluated in Section 4, also considering the influence of packet loss as well as false and missed sensor activations. We conclude the paper in Section 5 with a summary and directions for future work.

2   Related Work

Object tracking has often been investigated for military purposes or basic surveillance applications. In contrast to our work, various approaches assume sensors that generate signals dependent on the distance from a tracked object [3] or use geographical information [4]. However, we assume that devices are equipped with a sensor providing only binary signals (like passive infrared sensors) and consider an indoor scenario without location information. Therefore, we do not review these methods further.

Other approaches that deal with binary sensors [5,6,7] consider only a single target, assume rather dense networks [6,8], or reliable sensor readings [5]. All these approaches aim at finding an object's trajectory within a two-dimensional area using particle filters. Singh et. al. propose a one-dimensional tracking algorithm called *ClusterTrack* [9]. They also track multiple targets by applying particle filters. They focus on the effectivity of their scheme rather than on the used communication protocols and hence, propose a non-distributed scheme that assumes all sensor readings being available at a central processing unit which estimates the target's trajectories. As a result, the proposed scheme is not applicable to sensor networks without adaptation.

The authors of [10] present the restriction of particles to a graph and try to infer typical motion patterns of objects within a building. However, they require that objects can be identified (which limits their approach to a single object).

A different approach dealing with graphs is described in [11]. The authors formulate the tracking problem as a hidden state estimation problem in a Markov model and then derive their tracking scheme from the Viterbi algorithm [12].

Unfortunately, they assume perfect knowledge about the number of targets and their identities, which is not available in our scenario.

The idea of decomposing networks with holes into sections or clusters and representing them as a graph was first proposed in [1]. The authors then present a method for achieving that kind of topology recognition in [2]. It is particularly targeted at border and junction detection, but is designed for extremely high network densities. Hence, the approach is not directly applicable to the scenario considered here.

3   Decentralized Object Tracking

In this section we present our location-free and robust in-building object tracking scheme for wireless sensor networks. It can be divided into the start-up phase that is executed after network deployment and the subsequent operation phase.

The basic idea of our scheme is that each node pre-computes the one-dimensional spatial ordering of the nodes within its communication range during the start-up phase. These orderings are then regarded as the expected sensor activation patterns of neighbor nodes when an object passes by.

During operation, event sequences can be compared to these patterns. In addition, the event sequences are also stored in a short history. Like this, new patterns that result from network changes due to node failures or addition of nodes can be learned over time.

Our object tracking algorithm requires a loose time synchronization (approximately half a second as a maximum of time difference, depending on the maximum object velocity) to ensure that sensor activation events can be ordered correctly in time. Other protocols however, often impose more rigid time synchronization requirements such as duty-cycling or slotted medium access. Furthermore, we explicitly do not require that devices are location-aware. Instead, we use the distance estimation scheme described in [13]. It is based on neighborhood list comparisons instead of relying on Received Signal Strength Indication (RSSI) or Time Difference of Arrival (TDoA) measurements that are error-prone especially in indoor settings.

3.1   Start-Up Phase

The start-up phase is executed directly after network deployment. It can be repeated after times of high fluctuation or network restructuring, but this is usually not required as the algorithm works adaptively during the operating phase. During setup, the devices

1. calculate the local topology, i.e. the one-dimensional ordering of neighbor groups (usually one on each side),
2. find out whether they reside at corridor junctions or dead ends, and
3. assign corridor section memberships to their neighbors.

In the following, we discuss the start-up phase from the perspective of a node N (marked as a dark gray square in Figure 2(a)). At first, N detects the *one-dimensional ordering* of its neighbors. Details on this process are provided in [14], so we give only a short overview here. N broadcasts a 'hello' message to advertise its presence to its neighbors, and in turn collects its neighbors' messages to build up a neighborhood list.

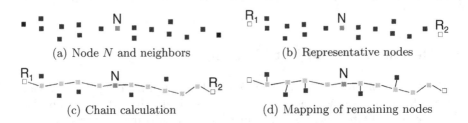

(a) Node N and neighbors   (b) Representative nodes

(c) Chain calculation   (d) Mapping of remaining nodes

Fig. 2. Local topology recognition

The resulting list is broadcasted again. When receiving such a list, it is compared with the local list of neighbors. A distance estimate can be derived from the fraction of common neighbors. Please refer to [13] for a detailed description of the distance estimation process. As a result, N can augment its local list with information on the distance to its neighbors.

The augmented list broadcasted again, resulting in all nodes knowing the distances of their neighbors from each other. N now sorts its neighbors by ascending distance and selects the furthest 40% of them. It then divides this subset into groups, so that neighbors that are close to each other are in same group whereas distant neighbors are in different groups. Following this, the most distant node from N in each group is chosen to be a group representative (white squares R_1 and R_2 in Figure 2(b)). Finally, so called 'chains', i.e. orderings of neighbors between the representatives and N are computed based on distance information between R_i and N (cf. Figure 2(b)). Nodes that are not part of the chains are mapped onto the nearest chain member (cf. Figure 2(d)).

Based on the number of chains, nodes can *decide whether they reside at junctions or dead ends*, i.e. whether they belong to the graph's vertices. In a perfect world most nodes would have ended up with two chains because they are situated somewhere in a corridor, 'vertex nodes' near junctions or dead ends would have calculated one, three or four chains. However, influences like distance estimation errors can lead to errors here. To increase robustness, a local voting process is employed: if a node's chain number does not equal two, it broadcasts it to its neighbors. Hence, nodes learn the chain counts of their neighbors. Nodes only consider themselves as vertex nodes if at least one third of their neighbors are also vertex nodes because they calculated the same number of chains. Again for each vertex node set, a representative is elected to find the optimally located node. Nodes then calculate the average distance to the other vertex nodes.

The one with the smallest average can be expected to reside in the center of the cluster, and hence becomes its representative.

Finally, all nodes determine their neighbors' *corridor membership*. By default, all neighbors are assumed to reside in the same corridor as oneself. However, vertex representatives broadcast their chains (including the mapped nodes) to their neighbors. If a node receives such a message, it checks which chain it belongs to, and marks all its neighbors that are not in the same chain to reside in a different corridor.

3.2 Operating Phase

After start-up, the network enters its regular operation mode, that is presented in this section. Remember that the core idea is to compare sensor activation patterns with the pre-computed chains, but that nodes should be enabled to adapt to new patterns over time. Therefore the chains constitute the foundation stone for a *history* of patterns that is amended during operation.

Whenever a node's sensor activates, it broadcasts a so-called *ObjectMessage* containing

- node ID: the ID of the activated node,
- activation time,
- starting location: the ID of the representative of last passed corridor vertex,
- starting time: the time when the object entered the current corridor section (Note that starting location and time identify an object uniquely because we assume that no two objects can activate the same sensor at exactly the same time),
- activation count: the number of activations since the object entered the current corridor section,
- velocity: average inter-activation time of the object.

Nodes store incoming activation messages in their *message buffer*. Messages in the buffer are sorted by the object identification, messages concerning the same object by the activation time. The message buffer has a fixed size where newer messages replace older ones. The buffer size should ensure that messages can remain in it for at least the time it takes at most for an object to cross the sensors' communication range.

Let us assume for example that node N_{15} received 4 object messages (two from N_{12} and one from N_{13} and N_{14}). Than, its message buffer could contain two objects:

$O_{id} = (S_{Loc}), S_{Time}$	List of neighbors' activations
$(1, 5.5)$	$(12, 7.6)$ $(13, 9.1)$ $(14, 10.8)$
$(1, 10.8)$	$(12, 11.6)$

When a node's sensor is activated, it tries to find a sequence of activation messages in its buffer that can be correlated with a history entry. For each message sets i.e. messages that concern the same object, it constructs the according activation pattern by extracting the node ID from each message and appends itself at the end. So, if N_{15}'s sensor was activated at 12.2 in our example, there would be two resulting patterns: 12, 13, 14 and 12.

The node then compares all these patterns to all history entries. The pattern that matches best is removed from the message buffer to conserve memory space. If exactly the same pattern already exists in the history, an associated match counter is increased; otherwise the pattern is inserted into the history. If no matching entry was found at all, a new one is created. Finally, an object message is constructed and broadcasted to the node's neighbors.

In the example let the (initial) history of N_{15} be

Match Counter	Last matching time	Pattern
0	0.0	12 13 14 15
0	0.0	18 17 16 15

Then, N_{15} finds the best matching by correlating the first message $((1, 5.5))$ with the first history entry. Hence, it increments the respective history entry counter, sets the last matching time to its activation time (12.2) and broadcasts its correlation result $((1, 5.5), (15, 12.2))$.

The core comparison mechanism is derived from the so-called *local alignment* [15] that was intended for finding similar sequences in DNA strings. It can tolerate entry permutations or missing entries to a certain degree, and delivers a degree of similarity. We adapted it slightly to our problem.

$$\text{Let } s : \mathbb{N}^n \times \mathbb{N}^m \rightarrow \mathbb{N}^{n \times m}, s_{i,j} = \begin{cases} 0 : a_i \neq b_j \\ 2 : a_i = b_j \wedge \|i - j\| > 2 \\ 4 : otherwise \end{cases}$$

be the equality matrix of two sequences $a = \{a_1, ..., a_n\}$ and $b = \{b_1, ..., b_m\}$, let d (with $d < 0$) be the penalty measure for a deletion and let F be the matrix for the computation of the local alignment with

$$F(0,0) = 0$$

$$F(i,j) = \max \begin{cases} F(i-1, j-1) + s(x_i, y_i) \\ F(i-1, j) + d \\ F(i, j-1) + d \\ 0 \end{cases}.$$

The matrix F is filled line-by-line, its maximum value $\max_{0 \leqslant i, j \leqslant m, n} F(i, j)$ indicates the degree of similarity, where higher values represent more similar patterns. As we do not consider dense networks, the quadratic time complexity should not be a problem. In particular is the size of the patterns limited to half of the number of neighbors. To avoid the quadratic space complexity the implementation uses only two vectors of size n for the computation. This is possible as it suffices to store the previous raw and the maximum value.

In addition to the local alignment, the comparison includes a number of sanity checks that can degrade the matching. They include the time since the last activation (derived from the message buffer), whether the velocity and last activation time fit the current activation and whether previous nodes reside in the same corridor. To adapt to new nodes or new orders of neighbors unknown patterns are added to the history and old history entries (depending on the last matching time) are removed from the history.

4 Evaluation

To evaluate our approach, we ran an extensive set of simulations with the scenario shown in Figure 1. The simulation area covers 50 by 50 meters. For all simulations 140 nodes equipped with a sensor monitoring its corridor sector over a length of 1 meter were placed on the wall of all corridors about every 1.5 meters.

The scenario includes four corridor segments, two junctions and one dead-end, allowing to evaluate a large number of different settings. In addition, two of the corridors feature two right angle corners. These areas are particularly difficult, as the distance estimation scheme tends to yield erroneous results here. Apart from wrong orderings, this may lead to false vertex detections.

We decided not to simulate the whole communication stack because we wanted to lay the focus of our work on effects that directly influence the object tracking rather then on networking aspects. For this reason we chose SHAWN [16] for our simulations. It models the effects of a MAC layer (e.g., packet loss, corruption and delay), but does not perform a complete simulation of the stack. It can simulate message exchange and measurements very efficiently and was developed for algorithmic simulations.

Instead of using the rather unrealistic unit disk graph radio propagation model, we employed the Radio Irregularity Model (RIM) [17]. It was developed to model radio properties of wireless sensor nodes and is a well established nowadays. Rather than using a fixed radio range, the maximum communication distance of a node also depends on the angle between sender and receiver. This in particular leads to a certain percentage of unidirectional links.

To achieve a reasonable network density, the average communication range was set to 9 meters. As a result, nodes on average had 6 neighbors on each side. While nodes within the corridor segments thus had about 12 neighbors, nodes had up to 20 neighbors at junctions and down to 6 neighbors in the dead-end. We consider this network density as a reasonable choice as wireless networks of smaller densities tend to partition [18,19].

We ran the object tracking scheme and analyzed the results of both start-up and operating phase. To obtain statistically sound results, we averaged the results of 100 simulations using the same parameter set but different random seeds.

4.1 Topology Recognition Evaluation

First, we evaluated the recognition of vertex nodes at junctions and dead-ends. For the subsequent object tracking, the absence of false vertex nodes within the

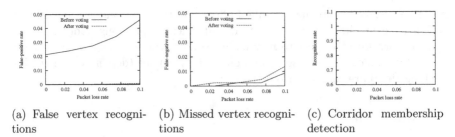

(a) False vertex recogni-   (b) Missed vertex recogni-   (c) Corridor membership
tions   tions   detection

Fig. 3. Quality of start-up phase

corridors is of the same importance as the presence of vertex nodes in junction areas.

Figure 3(a) shows the effectiveness of the voting process against false positives, i.e. nodes that reside in corridor segments but determined themselves as vertex nodes. While the pre-voting error rate rises from 2% to about 4.5% with increasing packet loss rate, the post-voting error rate is zero and literally does not increase at all.

As the voting successfully prevents false positives by reducing the number of vertex nodes, it must be ensured that it does not inhibit vertex detection at junctions.Figure 3(b) depicts the fraction of nodes that spuriously did not consider themselves as vertex nodes over different packet loss rates. While more or less all vertices are recognized for less than 7% packet loss, the rate of false negatives slightly rises for higher loss rates. However, it is also visible that the voting process has hardly any negative influence here. Nevertheless, it might be beneficial to resend messages to guarantee a good cluster recognition if extremely high packet loss rates are expected.

Finally, we evaluated the recognition of the corridor membership. Figure 3(c) indicates the fraction of neighbors that were assigned to the correct corridor section. It is obvious that only a small fraction of about 3% of the neighbors is miscategorized regardless of the packet loss rate. These errors are negligible for the object tracking scheme and can hence be tolerated.

4.2   Object Tracking Evaluation

We then evaluated the second part of our scheme, the operating phase.

We applied our scheme to various object movement patterns with different numbers of objects moving at different velocities. The patterns comprise objects passing by, following each other as well as overtaking each other.

In this section we present results for five selected movement patterns:

1. *1after2*: one object moves from C to A through corridors 4 and 1, then a second obejct takes the same way, and finally a third object moves the opposite way afterwards.
2. *Succeeding*: Two objects move from A to B through 1 and 4 following each other with a distance of 2 meters.

3. *Overtaking*: Two objects move from A to C through 1 and 4, one overtaking the other.
4. *1towards4*: Four objects move from A to C through 1 and 4 following each other with a distance of 5 meters, while at the same time a fifth object moves from C to A through the same corridor segments.
5. *3Ways*: Three objects move from A to C, one object through corridor 1, one through 2 and one through 3,at speeds that ensure their simultaneous arrival.

Influence of Activation Time Variation and Packet Loss. It must be assumed that not all sensors are adjusted perfectly with regard to orientation and position. To account for such effects, we added a random error to the activation times to obtain realistic simulations. Additionally, this compensates for the unnaturally perfect deployment of nodes in the simulations which may not be possible in real buildings.

In our simulation model the perfect activation occurs when the object reaches the point closest to the sensor (given that the object intrudes the detection range at all). We added a random offset to that point in time.

Precipitated and delayed activations cause a certain fraction of swapped activations which aggravates the object correlation significantly. In addition, the estimated velocity becomes extremely error-prone, impeding the differentiation between overtaking and closely succeeding objects.

We obtained the random error values from a Gaussian distribution with a standard deviation of $\sigma = 0.45$. Apart from the false activation time itself, this causes 1.7% of the activations to swap. Figure 4 and Figure 5 show the results of these simulations.

First of all one can see in Figure 4 that all objects are correctly tracked if neither the activation times are error-afflicted nor packets are lost. Correcty tracked means that for each corridor segment the object's start and end locations and times have been correctly recognized by a one vertex node. We then increased the packet loss rate up to 10%. The rate of correctly tracked objects decreases slowly for the more difficult movement patterns like *1towards4* and *3Ways*. Given a packet loss rate of 10% here, about 95% of the objects are correctly tracked. The moving pattern *3Ways* is difficult because of the objects simultaneously leaving from and arriving at a junction. If two objects arrive at a junction at the same time and a node's sensor detects the first one, the node must decide which previous activations in the neighborhood correlate better, but as both objects could continue their way, both objects are contemplable to be the reason for this activation.

As depicted in Figure 5 the results with the time-varied activations are nearly as good as the results without error. This shows the robustness of our tracking scheme.

Influence of Missed Activations. As hardware errors may lead to the oversight of moving objects by a sensor, we added activation miss rates of up to 10%. Figure 6 shows the result tracking rates without and Figure 7 shows result

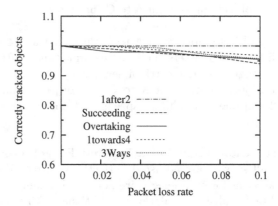

Fig. 4. Standard deviation $\sigma = 0$

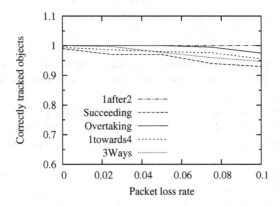

Fig. 5. Standard deviation $\sigma = 0.45$

tracking rates with the sensor activation time variance from the last section. For these simulations, the packet loss rate was set to 0.

For activation failure rates of up to 0.05 the results closely resemble those without sensor failure but with a corresponding packet loss rate. This is hardly surprising because for a node's neighbors it does not make any difference whether it failed to activate or whether the resulting packet was lost; it only make a difference for the directly affected node. If there are e.g., two objects closely succeeding each other and the sensor activates caused by the second object, having missed the first object may result in the wrong correlation decision which is not the case for packet loss. This difference is responsible for the slightly lower tracking rates at high failure rates compared to the results of the previous section.

However, experiments with real passive infrared sensors show that these tend to overreact rather than to miss events. Hence, we also evaluated influence of false positive activations in the following section.

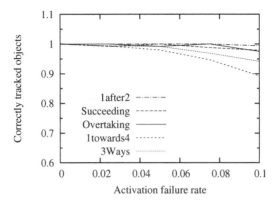

Fig. 6. Standard deviation $\sigma = 0$

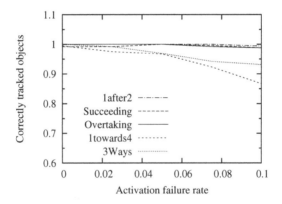

Fig. 7. Standard deviation $\sigma = 0.45$

Influence of False Positive Activations. Finally, we investigated the influence of additional false activations caused by hardware errors, temperature variation and air draft. We considered two cases where we generated 29 respectively 58 random additional activations per node and day. These rates correspond to about 15% and 30% of false activations given each node activates one time during each simulation.

Figure 8 shows the resulting rates of correctly tracked objects for different activation time variances and packet loss rates. It becomes clear that additional activations hardly influence the object tracking scheme: While the error increases by about 2% for low packet loss rates, the ghost activation influence even decreases for higher packet loss rates. If every tenth packet is lost, the false positive activations lose their influence completely.

However, it can be noted that different movement patterns suffer to a varying extent. Obviously, ghost activations have most influence when multiple objects are close to each other: The additional activations make it more difficult to assign sensor activations to the correct objects.

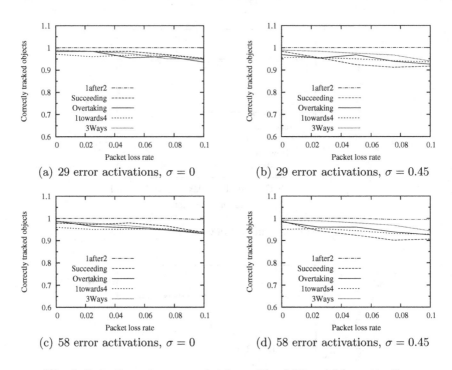

(a) 29 error activations, $\sigma = 0$ (b) 29 error activations, $\sigma = 0.45$

(c) 58 error activations, $\sigma = 0$ (d) 58 error activations, $\sigma = 0.45$

Fig. 8. Detection rate over packet loss with additional false activations

5   Conclusion

In this paper we presented a new approach for location-free object tracking in complex network topologies. Its core idea is to decompose the network into different areas where the tracking problem can be reduced to one dimension. Consequently, it becomes possible to do object tracking without location-aware nodes that are required by most other schemes.

By running an extensive set of simulations we showed that our scheme can not only handle a wide range of complex movement patterns, but is also robust against packet loss and sensor failures.

As future work, we plan to implement the algorithm on real hardware to enable further evaluation under real-life conditions.

References

1. Buschmann, C., Fekete, S.P., Fischer, S., Kröller, A., Pfisterer, D.: Koordinaten-freies Lokationsbewusstsein (Localization without Coordinates). IT - Information Technology, Themenheft Sensornetze 47(4) (April 2005)
2. Kröller, A., Fekete, S.P., Pfisterer, D., Fischer, S.: Deterministic boundary recognition and topology extraction for large sensor networks. In: Seventeenth Annual ACM-SIAM Symposium on Discrete Algorithms (SODA 2006) (2006)

3. Chellappa, R., Qian, G., Zheng, Q.: Vehicle detection and tracking using acoustic and video sensors. In: 2004 IEEE International Conference on Acoustics, Speech, and Signal Processing, May 17-21, 2004, vol. 3, pp. 793–796. IEEE, Los Alamitos (2004)
4. Pahalawatta, P.V., Depalov, D., Pappas, T.N., Katsaggelos, A.K.: Detection, classification, and collaborative tracking of multiple targets using video sensors. In: Zhao, F., Guibas, L.J. (eds.) IPSN 2003. LNCS, vol. 2634, pp. 529–544. Springer, Heidelberg (2003)
5. Aslam, J.A., Butler, Z.J., Constantin, F., Crespi, V., Cybenko, G., Rus, D.: Tracking a moving object with a binary sensor network. In: Akyildiz, I.F., Estrin, D., Culler, D.E., Srivastava, M.B. (eds.) SenSys., pp. 150–161. ACM, New York (2003)
6. Kim, W., Mechitov, K., Choi, J.Y., Ham, S.K.: On target tracking with binary proximity sensors. In: IPSN, pp. 301–308. IEEE, Los Alamitos (2005)
7. Shrivastava, N., Mudumbai, R., Madhow, U., Suri, S.: Target tracking with binary proximity sensors: fundamental limits, minimal descriptions, and algorithms. In: Campbell, A.T., Bonnet, P., Heidemann, J.S. (eds.) SenSys, pp. 251–264. ACM, New York (2006)
8. Arora, A., Dutta, P., Bapat, S., Kulathumani, V., Zhang, H., Naik, V., Mittal, V., Cao, H., Demirbas, M., Gouda, M.G., ri Choi, Y., Herman, T., Kulkarni, S.S., Arumugam, U., Nesterenko, M., Vora, A., Miyashita, M.: A line in the sand: a wireless sensor network for target detection, classification, and tracking. Computer Networks 46(5), 605–634 (2004)
9. Singh, J., Madhow, U., Kumar, R., Suri, S., Cagley, R.: Tracking multiple targets using binary proximity sensors. In: Abdelzaher, T.F., Guibas, L.J., Welsh, M. (eds.) IPSN, pp. 529–538. ACM, New York (2007)
10. Liao, L., Fox, D., Hightower, J., Kautz, H., Schulz, D.: Voronoi tracking: Location estimation using sparse and noisy sensor data (2003)
11. Oh, S., Sastry, S.: Tracking on a graph. In: IPSN, pp. 195–202. IEEE, Los Alamitos (2005)
12. Viterbi, A.: Error bounds for convolutional codes and an asymptotically optimum decoding algorithm. IEEE Transactions on Information Theory 13(2), 260–296 (1967)
13. Buschmann, C., Hellbrück, H., Fischer, S., Kröller, A., Fekete, S.: Radio propagation-aware distance estimation based on neighborhood comparison. In: Proceedings of the 14th European conference on Wireless Sensor Networks (EWSN 2007), Delft, The Netherlands (January 2007)
14. Blinded
15. Setubal, J., Meidanis, J.: Introduction to computational molecular biology. PWS Publishing Company, Setubal (1997)
16. Kröller, A., Pfisterer, D., Buschmann, C., Fekete, S.P., Fischer, S.: Shawn: A new approach to simulating wireless sensor networks. In: Design, Analysis, and Simulation of Distributed Systems 2005, Part of the SpringSim 2005 (April 2005)
17. Zhou, G., He, T., Krishnamurthy, S., Stankovic, J.A.: Impact of radio irregularity on wireless sensor networks. In: MobiSys 2004: Proceedings of the 2nd international conference on Mobile systems, applications, and services, pp. 125–138. ACM Press, New York (2004)
18. Hellbrück, H., Fischer, S.: Towards analysis and simulation of ad-hoc networks. In: Proceedings of the 2002 International Conference on Wireless Networks (ICWN 2002), Las Vegas, Nevada, USA, pp. 69–75. IEEE Computer Society Press, Los Alamitos (2002)
19. Xue, F., Kumar, P.R.: The number of neighbours needed for connectivity of wireless networks. IEEE Wireless Networks 10(2), 169–181 (2004)

Reasoning about Context in Uncertain Pervasive Computing Environments

Pari Delir Haghighi, Shonali Krishnaswamy, Arkady Zaslavsky,
and Mohamed Medhat Gaber

Center for Distributed Systems and Software Engineering
Monash University, Australia
{pari.delirhaghighi,shonali.krishnaswamy,arkady.zaslavsky,
mohamed.gaber}@infotech.monash.edu.au

Abstract. Context-awareness is a key to enabling intelligent adaptation in pervasive computing applications that need to cope with dynamic and uncertain environments. Addressing uncertainty is one of the major issues in context-based situation modeling and reasoning approaches. Uncertainty can be caused by inaccuracy, ambiguity or incompleteness of sensed context. However, there is another aspect of uncertainty that is associated with human concepts and real-world situations. In this paper we propose and validate a Fuzzy Situation Inference (FSI) technique that is able to represent uncertain situations and reflect delta changes of context in the situation inference results. The FSI model integrates fuzzy logic principles into the Context Spaces (CS) model, a formal and general context reasoning and modeling technique for pervasive computing environments. The strengths of fuzzy logic for modeling and reasoning of imperfect context and vague situations are combined with the CS model's underlying theoretical basis for supporting context-aware pervasive computing scenarios. An implementation and evaluation of the FSI model are presented to highlight the benefits of the FSI technique for context reasoning under uncertainty.

Keywords: context, fuzzy logic and pervasive computing.

1 Introduction

In pervasive computing environments, applications need to be aware of the changes in their environment and adapt their behavior according to these changes. Pervasive systems use context-awareness to perform their tasks in an intelligent and efficient manner and maintain consistency and continuity of their operations. Context is a very broad term that encompasses different aspects and characteristics [1]. Context can be related to a network, application, environment, process, user or device. Contextual information collected from every single sensor or data source represents a partial view of the real-world. Aggregation of data from multiple sensors and sources provides a wider and more general view of surrounding environment and situations of interest [2]. For example, in a smart room scenario, rather than monitoring sensed context from light, noise and motion sensors individually, this information can be used to reason about situations such as 'meeting', 'presentation' or 'study' which provides a

D. Roggen et al. (Eds.): EuroSSC 2008, LNCS 5279, pp. 112–125, 2008.

better understanding of the environment. As a meta-level concept over context, we define the notion of a situation that is inferred from contextual information [2]. Situation-awareness provides applications with a more abstract view of their environment rather than focusing on individual pieces of context.

One of the main challenges in enabling situation-awareness in pervasive applications is managing uncertainty. Uncertainty can be related to context imperfection such as sensors' inaccuracy, missing information or imperfect observations [3-4]. However, there is another dimension of uncertainty that is inherent in human concepts and every day situations. In real-world, situations evolve and change into other situations (e.g. 'walking' changes to 'running'). Changes that occur between situations of 'walking' and 'running' are also good indicators of situations that may emerge – albeit with some vagueness and uncertainty. These uncertain situations can be of high importance to certain applications such as a health monitoring application that needs to monitor details of changes in a patient's health situation. To model real-world situations, reasoning approaches need to be able to reflect this aspect of uncertainty in the situation reasoning results.

Reviewing recent works [5-14] in context reasoning under uncertainty reveals that these works have limited capability in dealing with vagueness of real-life situations and reflecting gradual and delta changes in the results of situation inference. More importantly, they lack a rich theoretical basic for supporting pervasive computing scenarios. A formal and general context modeling and reasoning approach that is specifically developed for context-aware computing environments and can deal with uncertain context and vague situations is still an open issue in this area of research.

In this paper we present a novel approach called Fuzzy Situation Inference (FSI) for situation modeling and reasoning under uncertainty. The FSI model integrates fuzzy logic principles into the Context Spaces (CS) model [2], a theoretical approach for modeling context and situations. The CS model provides heuristically-based sensor data fusion algorithm, specifically developed for pervasive computing environments to deal with inaccuracies of sensory originated information (i.e. reliability and error of reading) and characteristics of context [15-16]. The FSI technique incorporates the CS model's underlying theoretical basis for supporting context-aware and pervasive computing environments while using fuzzy logic to model and reason about vague and uncertain situations.

This paper is structured as follows. Section 2 reviews the current state-of-the-art in context modeling and reasoning under uncertainty. Section 3 briefly discusses the Context Spaces (CS) model. Section 4 describes integration of fuzzy logic into the CS model as the FSI model. Section 5 and 6 present the implementation and evaluation of the FSI model respectively. Section 7 concludes the paper and discusses future work.

2   Related Work

Situation modeling and reasoning can range from simple conditional rules to more complex techniques. In a simple and basic way, Goslar and Schill [17] model context and situations using Topic Maps or Context Maps that represent real-world objects as topics. Schillit [18] captures the situational context using vectors that describe "the condition of situation, the sensing device, the required accuracy and update rate". In an object-oriented way, CoCo [19] represents context and situations using a graphical language and abstractions such as class, object, scales and factory. In the Context Modeling Language (CML) model, Henricksen [20-21] defines situations using predicate logic. Predicates are evaluated against a set of variable bindings and a context but the results are restricted to 'true', 'false' or 'possibly true'.

Situations are high level context that are inferred from low level context based on rules or reasoning algorithms. One of the major challenges in the situation reasoning is dealing with uncertainty. Bayesian reasoning is one of the methods used for dealing with uncertainty. In [5], Bayesian technique is applied for location tracking where location is computed by integrating readings of inaccurate sensors and in [6] it is used for estimation of indoor locations of devices. The probability model proposed in [7] extends an ontology-based model that uses Bayesian networks to reason about uncertainty. Applying Bayesian reasoning has the limitation of knowing prior probabilities in advance and this knowledge might not always be available.

The Dempster-Shafer theory is a well-known technique used for addressing uncertainty in context-aware computing. In [8], a weighted Dempster-Shafer evidence combining rule is introduced based on the historically-estimated correctness rate of sensors. A different approach proposed in [9] applies the Demspter-Shafer algorithm for context reasoning and the rough set technique for context aggregation.

Compared with other reasoning methods, the use of multi-value logic is appealing feature of the fuzzy logic for modeling uncertainty. In [10], a fuzzy representation of context is introduced for adaptation of user interface application on mobile devices and the same fuzzy concept has been used in [11] for providing the user with an explicit and meaningful explanation for the system's proactive behavior. Alternatively, in [12-13] fuzzy logic is used for defining the 'context situations' and the rules for adaptation of the service policies according to their fitness degree. The concept of situational computing using fuzzy logic presented in [1] is based on pre-developed ontologies and a similarity-based situation reasoning. Ranganathan et al. in [14] apply probabilistic logic when there is precise knowledge of event probabilities and fuzzy logic when this knowledge is not available.

Review of context modeling and reasoning approaches shows that most of these works do not provide a general approach that can be applied to different domains and have limited support for context-aware pervasive computing scenarios. A formal and unified context modeling and reasoning approach that can address different aspects of uncertainty in pervasive computing environments has not been introduced in the current state-of-the-art. The next section discusses the Context Spaces (CS) model and its underlying concepts and introduces the heuristics of CS for context reasoning.

3 The Context Spaces (CS) Model

The CS model represents contextual information as geometrical objects in multidimensional space called situations [2]. The basic concepts of the CS model are the context attribute, application space, context state and situation space.

A 'context attribute' describes any data used in the situation reasoning. The term 'application space' defines the universe of discourse and 'context state' refers to a collection of context values in CS. The concept of a 'situation space' is characterized by a set of regions. Each 'region' is a set of acceptable values of a context attribute that satisfies a predicate. A region is a crisp or conventional set of context attribute values such that any element is its member or not.

For example, a situation space called 'healthy' can be defined with a context attribute of heart rate. The region of values of heart rate can be between 45 and 85 bpm that satisfy two predicates of ≥45 bpm and ≤85 bpm. A context state with the value of 78 for heart rate is contained in the situation space of healthy and a context state with the value of 104 is not contained in that situation space.

In addition to basic concepts and techniques for situation modeling and reasoning, the CS model provides heuristics developed specifically for addressing context-awareness under uncertainty. These heuristics are integrated into reasoning techniques that are utility-based data fusion algorithms and compute the confidence level in the occurrence of a situation [14-15]. The two main heuristics of the CS model are as follows:

1. Individual significance (i.e. weight) and contribution of context attributes in the situation space
2. Inaccuracies of sensory originated information

These two heuristics deal with importance of each context attribute and sensors' inaccuracies. To enable situation-awareness in pervasive applications, it is imperative to address the issue of uncertainty. The CS deals with uncertainty mainly associated with sensors' inaccuracies. Yet there is another aspect of uncertainty in human concepts and real-world situations that needs to be represented in a context model and reflected in the results of situation reasoning. Fuzzy logic has the benefit of representing this level of uncertainty using membership degree of values.

The next section introduces the FSI model and discusses the CS model's heuristics in more detail.

4   The Fuzzy Situation Inference (FSI)

The FSI model maps situation modeling concepts and reasoning methods of the CS model into a fuzzy structure and tailors them to conform to fuzzy logic principles. The following subsections discuss situation modeling and reasoning in the FSI model.

4.1   Modeling Situations

In the FSI model, the term linguistic variable is used to express a 'context attribute'. Unlike context attributes, values that linguistic variables take are not numeric and are called terms (also known as fuzzy variables) [22]. Each term of a linguistic variable represents a fuzzy set that takes a pair of numeric values (i.e. a value and its membership degree). In a fuzzy set, unlike a region, membership of an item is gradual and is represented by a membership degree between 0 and 1 [23-25].

Definitions of 'application space' and 'context state' are applied similarly to the FSI model but the 'situation space' is differently defined. In FSI, a situation is defined by a set of fuzzy sets that are expressed as a FSI rule. Unlike CS, a situation can also be defined using multiple rules that have dependent or overlapping conditions to provide more flexibility in representing situations.

A FSI rule consists of multiple conditions joined with the AND operator where each condition can itself be a disjunction of conditions (i.e. using the OR operator) [26]. Each condition tests the input value using a membership function that corresponds to a fuzzy term. The consequent of the rule represents the output that suggests the degree of confidence in the occurrence of a situation. If the output of a rule evaluation for the 'hypertension' situation yields the value of 0.885, we can suggest that the level of confidence in the occurrence of 'hypertension' is 0.885. This value can be compared to a confidence threshold ε between 0 and 1 (i.e. predefined by the application's designers) to determine whether a situation is occurring.

4.2 Situation Reasoning

The two reasoning methods of the CS model that we discuss here are based on the first and second heuristics introduced in Section 3.

Situation reasoning based on weights and contribution level. The first heuristic of CS deals with the weights of context attributes and the level of confidence of attributes' values. Weights are values between 0 and 1 that are assigned to context attributes and represent relative importance of each context attribute for inferring a situation. A level of confidence is assigned to each element and reflects how that element relates to the modeled situation. The reasoning computation method based on weights and contribution levels of elements is as follows.

$$\text{Confidence} = \sum_{i=1}^{n} w_i c_i .$$
(1)

where w_i presents the weight assigned to context attributes and c_i denotes the confidence level of a context attribute. The contribution function that assigns the confidence values is proposed at a conceptual level and its implementation is later introduced in the second reasoning method based on sensors' inaccuracy.

In FSI, the concept of weights is associated with the conditions of a rule but the concept of a contribution level is implemented in a different way. The FSI equivalent to the equation (1) is a rule evaluation method that computes a level of certainty between 0 and 1 using membership functions and presented as follows.

$$\text{Certainty} = \sum_{i=1}^{n} w_i \mu(x_i) .$$
(2)

where $\mu(x_i)$ denotes the membership degree of the element x_i and w_i represents a weight assigned to a condition. If the OR operator is used it will be evaluated using the maximum function. The result of $w_i \mu(x_i)$ represents a weighted membership degree of x_i and n represents the number of conditions in a rule ($1 \leq i \leq n$).

Situation reasoning based on sensors' inaccuracy. To provide automatic computation of the contribution level at run-time, the second reasoning technique of CS incorporates the heuristic of sensors' inaccuracy presented as follows.

$$\text{Confidence} = \sum_{i=1}^{n} w_i . \Pr(\hat{a}_i^t \in A_i) .$$
(3)

where $\Pr(\hat{a}_i^t \in A_i)$ presents the confidence level of a context attribute value by computing the probability of a context attribute correct value $\hat{a}_i^t$ being contained in the region A_i. To compute the probability value based on the reliability of a sensor, the reliability of reading (e.g. 95%) is used to represent the probability value (i.e. $\Pr(\hat{a}_i^t \in A_i)=0.95$). Second option to compute the probability value is to integrate the sensors' inaccuracy of reading rather than the reliability of reading. Using this option, the probability value is calculated in the following format:

$$\Pr(e_j \leq a_i^t - \min(A_i^j)) - \Pr(e_j \leq a_i^t - \max(A_i^j)). \tag{4}$$

where a_i^t denotes the sensed value of the context attribute, e_j denotes the sensor reading error (i.e. a_i^t - $\hat{a}_i^t$) and $\min(A_i^j)$ and $\max(A_i^j)$ represent minimum and maximum values of the region. This reasoning technique requires the estimation of the reading error distribution of sensors.

The CS equation (3) deals with uncertainty factoring in inaccuracies of sensors however this equation does not reflect delta changes of values in the equation and is not adequate to reason about vague situations. The FSI equivalent to the CS equation (3) not only incorporates the contribution level associated with sensors' inaccuracy but includes the membership of the values as another factor affecting the contribution level. In the FSI model, we first calculate the correct value based on the reliability or error rate and then pass it to the membership function as follows.

$$\text{Certainty} = \sum_{i=1}^{n} w_i \mu(f(x_i, e_i)) \tag{5}$$

where w_i represents a weight assigned to a condition and $\mu(f(x_i, e_i))$ denotes the membership degree of the element x_i. The function f calculates the correct value of the context based on the inaccuracy value e_i. If e_i is a reliability rate, the sensed value is multiplied by it and if it is an error rate (i.e. ±) it is added to the sensed value.

Although the CS model's heuristics and reasoning techniques deal with sensors' inaccuracy and characteristics of context attributes (i.e. not discussed in this paper), they are inadequate to represent the uncertainty associated with real-life and human concepts which tend to be abstract and imprecise.

The CS model computes a contribution level of context attribute values based on sensors' inaccuracy. This information might not be always available and obtainable but, more importantly, it is not sufficient for computing contribution levels of continuous values. This is due to the fact that there is an uncertainty factor related to the values that are near the boundaries of a region (i.e. maximum and minimum values). Using a fuzzy approach, this type of uncertainty can be represented and reflected in the situation reasoning results [3, 24, 25]. The next section presents the implementation of the FSI model.

5 Implementation

We have implemented a prototype of health monitoring application based on FSI in J2ME and deployed it on a Nokia N95 (shown in Fig.1). The prototype reasons about situations of 'normal', 'pre-hypotension', 'hypotension', 'pre-hypertension' and 'hypertension'. This application can be used by patients who suffer from blood pressure fluctuations. A trapezoidal membership function is used to compute membership degree of context values. Contextual information used for reasoning includes systolic and diastolic blood pressure (SBP and DBP) and heart rate (HR).

Fig. 1. The prototype of a FSI-based health monitoring application running on a Nokia N95 with an ECG biosensor

To capture the patient's heart rate, we have used a two lead ECG biosensor from Alive Technologies [27] that transmits ECG signals using Bluetooth to the mobile phone. For the blood pressure, we have used randomly generated data that simulates blood pressure fluctuations. The health monitoring application performs situation reasoning in real-time on the mobile device. Status bars on the mobile phone displays the level of certainty and confidence in the occurrence of each situation.

To evaluate the FSI model we have conducted a comparative evaluation of the FSI, CS and Dempster-Shafer techniques that is presented in the next section.

6 A Comparative Evaluation

To evaluate the FSI model, we have compared the FSI situation reasoning technique to the CS and Dempster-Shafer (hereafter DS) reasoning approaches. The purpose of this evaluation is first to validate the FSI model against a well-known reasoning technique such as DS and a context model developed for pervasive computing environments such as the CS model. The second objective of the evaluation is to highlight the benefits of the FSI for reasoning about uncertain situations.

In this evaluation, we have considered situations of 'hypotension', 'normal' and 'hypertension'. These situations are defined using context attributes of systolic blood pressure (SBP) with the scale of 40-170 mm Hg, diastolic blood pressure (DBP) with the scale of 20-150 mm Hg and heart rate (HR) with the range of 20-150 bpm.

Table 1 depicts modeling of the three situations in the CS model including the weights of attributes and their corresponding regions of values. Unlike FSI, the CS model uses crisp boundary for regions. To provide a similar and balanced range of data for evaluation of these approaches, the boundaries of regions are selected in a way that they match the values of fuzzy sets with membership degree of 0.5.

Table 1. Situation definitions in the CS model

Situation	Context attribute	Region of values	Weight
Hypotension	1=SBP	≤85	0.4
	2=DBP	≤60	0.4
	3=HR	≤45	0.2
Normal	1=SBP	>85 and ≤135	0.4
	2=DBP	>60 and ≤110	0.4
	3=HR	>45 and ≤85	0.2
Hypertension	1=SBP	>135	0.4
	2=DBP	>110	0.4
	3=HR	>85	0.2

Although FSI can represent a situation with multiple rules and each condition can be joined by the OR operator, we use one rule to define a situation and do not include the OR operator so that both models can be closely compared. The modeling of the three situations in the FSI model is presented in Table 2. Weights of conditions for the FSI rules conform to the weights specified for the context attributes in the CS model.

Table 2. Situation definitions in the FSI model

Situation	Linguistic Variable	Terms	Fuzzy set
represented below via rules	FSI 1=SBP 2=DBP 3=HR	low, normal, high low, normal, high slow, normal, fast	trapezoidal membership functions used

Rule1: if SBP is low and DBP is low and HR is low then situation is hypotension
Rule2: if SBP is normal and DBP is normal and HR is normal then situation is normal
Rule3: if SBP is high and DBP is high and HR is high then situation is hypertension

To apply the DS algorithm for reasoning about situations, we use the Dempster's rule of combination. The normalized version of the combination rule is as follows.

$$m(R) = \frac{\sum_{P \cap Q = R} m_i(P).m_j(Q)}{1 - \sum_{P \cap Q = \phi} m_i(P).m_j(Q)} \qquad (6)$$

where m(R) denotes the mass value computed for a proposition R given the evidences i and j. If R represents a situation, considering all existing propositions, the intersection of some of these propositions denoted as P and Q results in the proposition R (i.e. $P \cap Q = R$) and the intersection of other combinations of propositions results in an empty set.

To model the three situations of Hypotension (L), Normal (N) and Hypertension (H) with DS, we first need to define propositions and events. Since all three situations are incompatible we include a proposition of Unknown (U) that would consist of three situations. Then we identify the events and mass values that reflect the association of an event with the occurrences of each proposition. An example of the events and mass values are depicted in Table 3. Mass values are assigned in a way that they reflect to what degree each event indicates a situation.

Table 3. Definitions of events and mass values

Event	Mass values for Normal	Mass values for Hypotension	Mass values For Hypertension	Mass values for unknown	Total mass
SBPLow (40-85)	0	0.7	0	0.3	1
SBPMed(86-135)	0.7	0	0	0.3	1
SBPHigh(136-170)	0	0	0.7	0.3	1
DBPLow(20-60)	0	0.7	0	0.3	1
DBPMed(61-110)	0.7	0	0	0.3	1
DBPHigh(110-150)	0	0	0.7	0.3	1
HRSlow(20-45)	0.2	0.4	0	0.4	1
HRMed(46-85)	0.4	0.2	0.2	0.2	1
HRFast(86-150)	0.2	0	0.4	0.4	1

Since we have based our situations on three context attributes, we define three mass functions of m_1, m_2 and m_3 corresponding to each context attribute. Then we apply DS combination over all propositions and available evidence. For example, if we have the context values of 82 for SBP, 52 for DBP and 58 for HR, we combine evidence for the occurrence of hypotension (L) as follows.

$$m_{12}(L) = \frac{\sum_{Q \cap R = L} m_1(A).m_2(B)}{1 - \sum_{Q \cap R = \phi} m_1(A).m_2(B)} = \frac{m_1(L).m_2(L) + m_1(L).m_2(U) + m_1(U).m_2(L)}{1 - m_1(L).m_2(H) - m_1(H).m_2(L) - m_1(L).m_2(N) - m_1(N).m_2(L)} =$$

$$\frac{0.7 \cdot 0.7 + 0.7 \cdot 0.3 + 0.3 \cdot 0.7}{1 - 0.7 \cdot 0 - 0 \cdot 0.7 - 0.7 \cdot 0 - 0 \cdot 0.7} \approx 0.91$$

$$m_{12}(H) = \frac{\sum_{Q \cap R = H} m_1(A).m_2(B)}{1 - \sum_{Q \cap R = \phi} m_1(A).m_2(B)} = \frac{m_1(H).m_2(H) + m_1(H).m_2(U) + m_1(U).m_2(H)}{1 - m_1(H).m_2(L) - m_1(L).m_2(H) - m_1(H).m_2(N) - m_1(N).m(H)} =$$

$$\frac{0 \cdot 0 + 0 \cdot 0.3 + 0.3 \cdot 0}{1 - 0 \cdot 0.7 - 0.7 \cdot 0 - 0 \cdot 0 - 0 \cdot 0} = 0$$

$$m_{12}(N) = \frac{\sum_{Q \cap R = N} m_1(A).m_2(B)}{1 - \sum_{Q \cap R = \phi} m_1(A).m_2(B)} = \frac{m_1(N).m_2(N) + m_1(N).m_2(U) + m_1(U).m_2(N)}{1 - m_1(N).m_2(L) - m_1(L).m_2(N) - m_1(N).m_2(H) - m_1(H).m(N)} =$$

$$\frac{0 \cdot 0 + 0 \cdot 0.3 + 0.3 \cdot 0}{1 - 0 \cdot 0.7 - 0.7 \cdot 0 - 0 \cdot 0 - 0 \cdot 0} = 0$$

$$m_{12}(U) = \frac{\sum_{Q \cap R = U} m_1(A).m_2(B)}{1 - \sum_{Q \cap R = \phi} m_1(A).m_2(B)} =$$

$$\frac{m_1(U).m_2(U)}{1 - m_1(H).m_2(L) - m_1(L).m_2(H) - m_1(N).m_2(H) - m_1(H).m(N) - m_1(N).m_2(L) - m_1(L).m_2(N)} =$$

$$\frac{0.3 \cdot 0.3}{1 - 0 \cdot 0.7 - 0.7 \cdot 0 - 0 \cdot 0 - 0 \cdot 0 - 0.0.7 - 0.7.0} \approx 0.09$$

$$m_{123}(L) = \frac{\sum_{Q \cap R = L} m_1(A).m_2(B)}{1 - \sum_{Q \cap R = \phi} m_1(A).m_2(B)} = \frac{m_{12}(L).m_3(L) + m_{12}(L).m_3(U) + m_3(U).m_{12}(L)}{1 - m_{12}(L).m_3(H) - m_{12}(H).m_3(L) - m_{12}(L).m_3(N) - m_{12}(N).m_3(L)} =$$

$$\frac{0.91 \cdot 0.2 + 0.91 \cdot 0.2 + 0.09 \cdot 0.2}{1 - 0.91 \cdot 0.2 - 0 \cdot 0.2 - 0.91 \cdot 0.4 - 0 \cdot 0.2} = \frac{0.382}{0.454} \approx 0.841$$

The same DS reasoning computation presented above is used in our evaluation. Although, the DS theory has the strength of representing unknown or uncertainty, determination of mass values for propositions can be a difficult task, particularly that they can have impact on the other situations. For evaluation of CS and FSI, we use the equations (1) and (2) (discussed earlier). These techniques do not include the sensor's inaccuracy and could be compared to the DS method more accurately.

The dataset used for evaluation is generated continuously (data rate is 30 records/minute) in ascending order. For this set of experiments, we have used our data synthesizer to represent the different events defined in Table 3 that contribute to the occurrence of each pre-defined situation as well as the uncertain situations. Table 4 depicts a snapshot of 131 context states that is used along with their scales.

Table 4. The data used for the comparative evaluation

Context attribute scales	Corresponding DS events
SBP:40-65, DBP: 20-45, HR: 20-45	**SBPLow, DBPLow, HRSlow**
SBP:66-80, DBP: 46-60, HR: 46-60	SBPLow, DBPLow, HRMed
SBP:81-85, DBP: 61-65, HR: 61-65	SBPLow, DBPMed, HRMed
SBP:86-105, DBP: 66-85, HR: 66-85	**SBPMed, DBPMed, HRMed**
SBP:106-130, DBP: 86-110, HR: 86-110	SBPMed, DBPMed, HRHigh
SBP:131-135, DBP: 111-115, HR: 111-115	SBPLow, DBPHigh, HRHigh
SBP:136-170, DBP: 116-150, HR: 116-150	**SBPHigh, DBPHigh, HRHigh**

Fig. 2 presents the results of comparative evaluation of three reasoning approaches of CS, DS and FSI for situations of 'hypotension', 'normal' and 'hypertension'.

Fig. 2 shows three approaches of CS, DS and FSI have a relatively similar trend according to context changes. When the data corresponds to a pre-defined situation the results of three approaches almost overlap. This overlapping is more noticeable with the CS and FSI models as they are based on similar heuristics.

However, when changes of data indicate the occurrence of an unknown and uncertain situation, differences of reasoning results between CS, DS and FSI are more apparent. Compared to FSI, the results of situation reasoning by the CS and DS methods show sudden rises and falls with sharp edges when situations change which do not match the real-life situations. This is because the DS and CS approaches do not deal

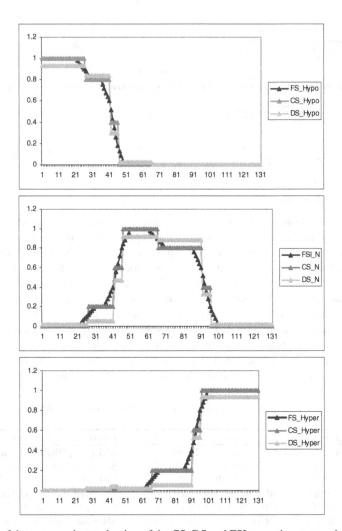

Fig. 2. Results of the comparative evaluation of the CS, DS and FSI reasoning approaches

with delta changes of the values and are not able to reflect the gradual evolution of one situation to another situation. When the value of context attributes decreases or increases, its membership degree also increases and decreases accordingly and gradually. This enables FSI to provide more accurate situation reasoning results in terms of reflecting very minor changes of context.

The evaluation validates the accuracy of the FSI model for situation modeling and reasoning and it also shows that FSI is able to reflect very minor changes of context in situation inference and represent changes in a more gradual and smooth manner.

The evaluation shows that the FSI model is more appropriate approach for representation of human concepts and for reasoning about the real-world situations that are defined by continuous values. Health-related situations are examples of these types of scenarios where FSI can prove to be more fitting approach compared to the DS and CS reasoning approaches.

7 Conclusion

Situation modeling and reasoning under uncertainty are challenging research tasks in context-aware pervasive computing. Fuzzy logic has the potential to represent the fuzziness and uncertainty that is associated with real-world situations. However, application of a fuzzy approach per se can not be adequate for situation modeling and reasoning in pervasive computing environments. Therefore, it is imperative that a fuzzy modeling and reasoning method to be combined with a rich theoretical basis for supporting context-aware scenarios.

In this paper, we proposed a Fuzzy Situation inference (FSI) model that integrates fuzzy logic into the CS model, a formal and general context reasoning and modeling technique for pervasive computing environments. The strengths of fuzzy logic for modeling imperfect context and reasoning about vague situations are combined with the CS model's underlying theoretical basis for supporting context-aware and pervasive computing scenarios. An implementation and evaluation of the FSI model were presented through a scenario in health monitoring to highlight the benefits of the FSI technique for context reasoning under uncertainty.

The FSI model is a part of our architecture for adaptive mobile data stream mining. In this project, we use the results of FSI for gradual tuning of parameters of data stream mining algorithms and perform intelligent and real-time analysis of data stream generated from sensors on mobile devices. The analysis is underpinned using situation-aware adaptation. In the future, we intend to explore and model relationships between situations, and extend FSI with learning capabilities so the system can predict situations.

References

1. Haghighi, P., Zaslavsky, A., Krishnaswamy, S.: An Evaluatation of Query Languages for Context-Aware Computing. In: The 1st International Workshop on Flexible Database and Information Systems Technology (FlexDBIST 2006), Held in conjunction with DEXA 2006 International Conference on Database and Expert Systems Applications. IEEE Computer Society Press, Crakow (2006)
2. Padovitz, A., Loke, S., Zaslavsky, A.: Towards a Theory of Context Spaces. In: Proceedings of the 2nd IEEE Annual Conference on Pervasive Computing and Communications, Workshop on Context Modeling and Reasoning (CoMoRea). IEEE Computer Society, Orlando (2004)
3. Anagnostopoulos, C.B., Ntarladimas, Y., Hadjiefthymiades, S.: Situational Computing: An Innovative Architecture with Imprecise Reasoning. The Journal of Systems and Software 80, 1993–2014 (2007)
4. Satyanarayanan, M.: Coping with Uncertainty, IEEE CS Pervasive computing. Journal, Modeling uncertainty in context-aware computing (2001)
5. Fox, D., Hightower, J., Liao, L., Schulz, D., Borriello, G.: Bayesian filtering for location estimation. IEEE Pervasive Computing (2003)

6. Castro, P., Munz, R.: Managing context data for smart spaces. IEEE Personal Communications 7(5), 4–46 (2000)
7. Gu, T., Pung, H., Zhang, D.: A Bayesian approach for dealing with uncertain contexts. In: The Proceeding of the Second International Conference on Pervasive Computing (2004)
8. Wu, H., Siegel, M., Stiefelhagen, R., Yang, J.: Sensor Fusion Using Dempster-Shafer Theory. In: Proc. of IMTC 2002, Anchorage, AK, USA (2002)
9. Jian, Z., Yinong, L., Yang, J., Ping, Z.: A Context-Aware Infrastructure with Reasoning Mechanism and Aggregating Mechanism for Pervasive Computing Application. In: Proceedings of the 65th IEEE Vehicular Technology Conference (VTC Spring 2007), Dublin, Ireland, pp. 257–261 (2007)
10. Mäntyjärvi, J., Seppanen, T.: Adapting Applications in Mobile Terminals Using Fuzzy Context Information. In: The Proceedings of 4th International Symposium on Mobile HCI 2002, Italy, pp. 95–107 (2002)
11. Byun, H., Keith, C.: Supporting Proactive 'Intelligent' Behaviour: the Problem of Uncertainty. In: Proceedings of the UM 2003 Workshop on User Modeling for Ubiquitous Computing, Johnstown, PA, pp. 17–25 (2003)
12. Cao, J., Xing, N., Chan, A., Feng, Y., Jin, B.: Service Adaptation Using Fuzzy Theory in Context-aware Mobile Computing Middleware. In: Proceedings of the 11th IEEE Conference on Embedded and Real-time Computing Systems and Applications (RTCSA 2005) (2005)
13. Cheung, R.: An Adaptive Middleware Infrastructure Incorporating Fuzzy Logic for Mobile computing. In: Proceedings of the International Conference on Next Generation Web Services Practices (NWeSP 2005) (2005)
14. Ranganathan, A., Al-Muhtadi, J., Campbell, R.H.: Reasoning about Uncertain Contexts in Pervasive Computing Environments. IEEE Pervasive Computing 3(2), 62–70 (2004)
15. Padovitz, A., Loke, S.W., Zaslavsky, A., Burg, B., Bartolini, C.: An Approach to Data Fusion for Context-Awareness. In: Dey, A.K., Kokinov, B., Leake, D.B., Turner, R. (eds.) CONTEXT 2005. LNCS (LNAI), vol. 3554, pp. 353–367. Springer, Heidelberg (2005)
16. Padovitz, A., Zaslavsky, A., Loke, S.W.: A Unifying Model for Representing and Reasoning About Context under Uncertainty. In: 11th International Conference on Information Processing and Management of Uncertainty in Knowledge-Based Systems (IPMU), Paris, France (July 2006)
17. Goslar, K., Schill, A.: Modeling Contextual Information Using Active Data Structures. In: Workshop for Pervasive Information Management (PIM), International Conference on Extending Database Technology (EDBT), Heraklion, Crete, Greece (2004)
18. Schilit, B.N., Theimer, M.M., Welch, B.B.: Customizing Mobile Applications. In: Proceedings USENIX Symposium on Mobile and Location-Independent Computing (USENJX Association) (1993)
19. Buchholz, T., Krause, M., Linnhoff-Popien, C., Schiffers, M.: CoCo: dynamic composition of context information. In: The 1st Annual International Conference on Mobile and Ubiquitous Systems: Networking and Services (MOBIQUITOUS 2004), Boston, Massachusetts (2004)
20. Henricksen, K., Indulska, J.: Modelling and Using Imperfect Context Information. In: Proceedings of the 2nd IEEE Annual Conference on Pervasive Computing and Communications. Workshop on Context Modelling and Reasoning (CoMoRea 2004). IEEE Computer Society, Orlando (2004)
21. McFadden, T., Henricksen, K., Indulska, J.: Automating context-aware application development. In: UbiComp 1st International Workshop on Advanced Context Modelling, Reasoning and Management, Nottingham, pp. 90–95 (2004)

22. Zadeh, L.A.: The Concept of a Linguistic Variable and Its Application to Approximate Reasoning Information Systems, pp. 199–249 (1975)
23. Mendel, J.M.: Fuzzy Logic Systems for Engineering: A Tutorial. Proceedings of the IEEE 83(3), 345–377 (1995)
24. Jang, J.R., Sun, C., Mizutani, E.: Neuro-Fuzzy and Soft Computing: A Computational Approach to Learning and Machine Intelligence. Prentice-Hall, Upper Saddle River (1997)
25. Zimmermann, H.J.: Fuzzy Set Theory - and Its Applications. Kluwer Academic Publishers, Norwell (1996)
26. Bruce, G., Buchanan, B.G., Shortliffe, E.D.: Rule-based expert systems: the MYCIN experiments of the Stanford Heuristic Programming Project. Addison-Wesley, Reading (1984)
27. Alive Technologies, http://www.alivetec.com

Contextual Ranking of Database Querying Results: A Statistical Approach

Xiang Li, Ling Feng, and Lizhu Zhou

Department of Computer Science and Technology,
Tsinghua University, Beijing, 100084, China
li-xiang06@mails.tsinghua.edu.cn
{fengling,dcszlz}@tsinghua.edu.cn

Abstract. There has been an increasing interest in context-awareness and preferences for database querying. Ranking of database query results under different contexts is an effective approach to provide the most relevant information to the right users. By applying the regression models developed in the statistics field, we present a quantitative way to measure the impact of context upon database query results by means of contextual ranking functions with context attributes and their influential database attributes as parameters. To make the approach computationally efficient, we furthermore propose to reduce the dimensionality of context space, which can not only increase computational efficiency but also help ones identify informative association patterns among context attributes and database attributes. Our experimental study on both synthetic and real data verifies the efficiency and effectiveness of our methods.

1 Introduction

With the proliferation of ubiquitous computing which integrates information processing with everyday activities, it is indispensable to provide effective data management support which is affected by preferences and needs of different users under different contexts. Therefore, context-aware preference querying techniques are in great demand. At the same time, the increasing amount of data stored in databases often leads to the *many-answers* problem. Contextual ranking of database query results according to how well they match user's preferences becomes an effective approach to make the right information available to the right user under the right context. We use the following running example throughout the paper to illustrate key concepts:

Example 1. *Table 1 shows a sun protection product relation with attributes TupleId, Name, Sun-Protection-Factor (SPF), Net Weight, and Price, and Table 2 shows some user-related context information, including monthly income and Ultra Violet Index (UVI) of the place where the user is located, together with some ranking scores that different query users assigned to different products. For instance, when a user is at a place, say Honolulu, with a high UVI, s/he will highly rank the sun protection cream with a high SPF.*

In the literature, two lines of research were conducted for database preference querying, namely, quantitative and qualitative [8]. The qualitative approach is to directly specify

D. Roggen et al. (Eds.): EuroSSC 2008, LNCS 5279, pp. 126–139, 2008.

Table 1. The *Product* relation

TupleID	Name	SPF	NetWT(ml)	Price(USD)
t_1	EL Cyber White	50	50	47
t_2	AVON SUN	40	50	9.99
t_3	Biotherm Sunfitness	15	125	21.50
t_4	CD Bronze	15	150	26
t_5	Sisley Sunleya	15	50	180

Table 2. Contextual ranking of *Product* tuples

ContextInstanceID	Income	UVI	t_1	t_2	t_3	t_4	t_5
u_1	5000	10 (Honolulu)	0.190	0.100	0.088	0.092	0.050
u_2	3000	2 (Seattle)	0.362	0.259	0.305	0.353	0.064
u_3	8000	1 (Chicago)	0.150	0.107	0.138	0.147	0.099
u_4	6500	7 (Miami)	0.228	0.110	0.096	0.100	0.062

preferences on database tuples, typically using binary preference relations and/or rules. A preference relation example is "*prefer one book tuple to another if and only if their IS-BNs are the same and the price of the first is lower.*" The quantitative approach expresses preferences using scoring functions, which associate numeric scores with database tuples. Tuple t_1 is said to be preferred to tuple t_2 if and only if the score of t_1 is higher than the score of t_2.

In [4], van Bunningen *et al.* took a qualitative approach to model users' context-aware preference rules using Description Logics. "*Eric prefers to watch the TV program on Channel 5 during weekends*" is an example of such a context-aware preference rule. While such a knowledge-based approach well suits querying reasoning and inference, it cannot measure the impact of context upon database query results in a relatively precise and quantitative way. In real world applications, the number of users' preference rules could be so large that figuring out the complete set of rules is usually impossible. Particularly when the rule set is large, confliction may easily emerge from logic rules. The aim of this study is to overcome the shortcomings of the qualitative approach, and investigate the impact of context, as measured in a more general quantitative way, on database querying, with a hope that the combination of qualitative and quantitative methods could complement each other to deliver a more satisfactory solution to context-aware database querying.

This paper reports our preliminary study on contextual ranking of database query results by exploring statistic correlations between context attributes and database attributes instead of generating hard-coded preference rules. We apply the regression models well-developed in the statistic field to approximate contextual ranking functions on context attributes and database attributes. As the context space encompasses multiple dimensions about users and external environments, data collection in the context database usually leads to an information overload. This enormity inevitably brings serious computational problems to the statistic relationship investigation between database tuples and context attributes. Therefore, we perform context selection to remove irrelevant and/or redundant context dimensions in order to improve the efficiency of ranking functions' learning and enhance the comprehensibility of learning results. We expect a small subset of context attributes which gives adequate measurement accuracy with a reasonable computational cost. On the other hand, in trying to understand the effect of context on database tuples, it may be desirable to consider context attributes which are either known or believed to affect the ranking scores. Empirical experiments are conducted to compare the performance of statistic regression models for context-aware query and demonstrate necessity and effectiveness of our context selection strategy.

The contribution of the paper is two-fold. First, we present a quantitative mechanism to measure the impact of context upon database query results by means of contextual ranking functions with context attributes and database attributes as parameters. Second, we present a method of reducing dimensionality of context space which not only increases computational efficiency but also helps us identify informative patterns among context attributes and database attributes, that is, which context elements influence the query answer. Dimension reduction is extremely important because there are always unavoidable gaps between tools and representations. Our context selection method could be considered as an instrument bridging the gaps.

The rest of this paper is organized as follows. We review some closely related work in Section 2. Contextual ranking of database query results is discussed in Section 3. We describe techniques for context selection in Section 4. In Section 5, we present our evaluation results on both synthetic and real data sets. Finally, Section 6 concludes the paper.

2   Related Work

In this section, we review two pieces of work which are closely related to this study.

Context-Aware Preference Querying. Users' context-aware preferences over database tuples can be expressed through either *quantitative* scores bound to interesting tuples, or *qualitative* relations/rules to be satisfied by interesting tuples, taking context information into account. For example, Stefanidis *et al.* used a combination of context and non-context attribute-value pairs associated with a degree of interest to define a context-aware preference [22]. They gave a hierarchical model for expressing contextual preferences [23], and proposed to locate and score database tuples corresponding to the given context descriptor. In a recent work, they grouped context-aware preferences based on score similarity, where similarity for each of context parameters is defined according to the path distance and the depth of the hierarchy levels. The distance between the context states is defined as the weighted sum of value distances [21].

Without considering context, Agrawal and Wimmers drew a general framework for quantitatively expressing and combining multiple preference functions [2]. Kießling [15], Koutrika and Ioannidis [16] presented rich preference models which associate degrees of interest with preferences over a database schema.

Recently, learning to rank starts to receive attention in the web search domain, where Chen *et al.* [7] proposed a two-phased framework for training ranking functions. First, manually labeled examples are used to train a base linear model. Second, user preference examples are used to tune the base model parameter. They derived user preferences from "clickthrough data" which is the query log of the search engine in connection with the log of links the users clicked [14].

In comparison, Agrawal *et al.* adopted a qualitative approach to present and utilize preference relationships among tuples to pre-compute tuple orderings and use them to provide ranked query answers [1]. van Bunningen *et al.* illustrated a knowledge-based way to model context-aware preference rules, where both contexts and preferences are treated in a uniform way using Description Logics [4]. Database query results are ranked according to the weights of preference rules and degrees of context uncertainty [5].

The work reported in this paper distinguishes from the previous ones in the following two aspects. First, we take a quantitative approach, and exploit the regression models well-developed in the statistics field to precisely measure the impact of context-awareness upon database querying by learning contextual ranking functions on context and database attributes. Second, to increase computational efficiency and meanwhile target at more informative patterns, we propose a method to reduce high-dimensional context space prior to function learning and invoking for query result ranking.

Dimension Reduction. Dimension reduction and feature selection play a pivotal roles in statistics, machine learning, and pattern recognition. Traditional algorithms in machine learning and data mining suffer from the *curse of dimensionality* [3], which refers to the very slow convergence rate of any estimator to the true value of a multivariate smooth function defined on a high dimensional space. Liu and Motoda offers an excellent overview of various feature selection techniques [18].

Principal Component Analysis is a classical statistical method for transforming and re-expressing multivariate data which seeks to identify and understand patterns of association across variables [17]. It is an unsupervised technique which does not include label information of data, by contrast *Linear Discriminant Analysis* uses the label information to discover an informative projection on which the two classes are better separated [9]. *Spectral Regression* proposed in [6] is based on regression and spectral graph analysis and can be performed in a supervised/supervised/semi-supervised situation.

3   Conceptual Ranking of Database Query Results

In comparison with traditional non-context-aware relational database querying whose query result is a set of tuples satisfying the query condition, the answer to a context-aware database query is a ranked list of database tuples, whose ordering is based on how well the user may prefer the tuples, taking both the query condition and the query context into account.

Formally, let $R(A_1, A_2, \ldots, A_m)$ be a relation schema containing m attributes $A = \{A_1, A_2, \ldots, A_m\}$, whose domains are denoted as $Dom(A_1), Dom(A_2), \ldots, Dom(A_m)$, respectively. A relation of the relation schema $R(A_1, A_2, \ldots, A_m)$, denoted by $r(R)$, is a set of tuples $r(R) = \{t_1, t_2, \ldots, t_s\}$. Each m-tuple t is an ordered list of m values $t = \langle t.A_1, t.A_2, \ldots, t.A_m \rangle$, where each value $t.A_i$, $1 \leq i \leq m$, is an element of $Dom(A_i)$. In a similar fashion, an n-dimensional context space $CS(C_1, C_2, \ldots, C_n)$ contains n context attributes $C = \{C_1, C_2, \ldots, C_n\}$, whose domains are denoted as $Dom(C_1), Dom(C_2), \ldots, Dom(C_n)$, respectively. A context of the context space, denoted by $c(CS)$, consists of a set of context instances $c(CS) =$

$\{u_1, u_2, \ldots, u_s\}$. Each n-tuple u is an ordered list of n values $u = \langle u.C_1, u.C_2, \ldots, u.C_n \rangle$, where each value $u.C_j$, $1 \leq j \leq n$, is an element of $Dom(C_j)$.

In this study, we distinguish database attributes from context attributes in that: database attributes (like $Price$ and $NetWT$) are subject-oriented and inherent characteristics of database entity $product$; while context attributes (like $Income$ and UVI) are more user/environment-related whose values vary under different query contexts.

Definition 1. *Let $t \in r(R)$ be a tuple of relation $r(R)$, and let $u \in c(CS)$ be a context instance of context $c(CS)$. A **contextual ranking of database tuple** t under u is defined as a function: $f(t.A_1, t.A_2, \ldots, t.A_m, u.C1_1, u.C_2, \ldots, u.C_n) = s \in [0, 1]$.*

Definition 2. *Let $t, t' \in r(R)$ be two tuples of relation $r(R)$, and let $u \in c(CS)$ be a context instance of context $c(CS)$. We call t is **preferred to** t' under u, if and only if*

$$f(t.A_1, \ldots, t.A_m, u.C_1, \ldots, u.C_n) \geq f(t'.A_1, \ldots, t'.A_m, u.C_1, \ldots, u.C_n)$$

With the contextual ranking function, we can rank the answer to a context-aware database querying accordingly.

Definition 3. *Given a database query q over relation $r(R)$ under context instance $u \in c(CS)$, let $T_q \subseteq \{t_1, t_2, \ldots, t_q\}$ be a set of tuples satisfying the query condition of q without loss of generality. The **contextual ranking of query answer** T_q is defined as a function:*

$$CQ(T_q, u) = \{(t_i, s_i) \mid (t_i \in T_q) \wedge (s_i = f(t_i.A_1, \ldots, t_i.A_m, u.C_1, \ldots, u.C_n))\}$$

Example 2. *Suppose a query is issued under context instance u_3 to request sun products where $(NetWT < 100)$ (Table 1) with $\{t_1, t_2, t_5\}$ as the result tuples. According to Table 2,*
$f(t_1.SPF = 50, t_1.Price = 47, u_3.Income = 8000, u_3.UVI = 1) = 0.150$,
$f(t_2.SPF = 40, t_2.Price = 9.99, u_3.Income = 8000, u_3.UVI = 1) = 0.107$, and
$f(t_5.SPF = 150, t_5.Price = 180, u_3.Income = 8000, u_3.UVI = 1) = 0.099$.
Thus, we have $CQ(T_q, u_3) = \{(t_1, 0.150), (t_2, 0.107), (t_5, 0.099)\}$ as the final ranked query answer.

The establishment of an appropriate ranking function relies on users' subjective knowledge within a specific problem domain. Hence, the method proposed in this paper needs users to provide their preferences (measured in scores) to database tuples under various contexts. We use these information to learn the scoring function f, which is later used to predict users' inclinations to tuples under future (un)known contexts.

Now we can generalize our problem to adequately approximate a function of several to many variables given only the values of the function at various points in a dependent variable space. The goal is to model the dependence of a response variable score y on predictor variables $x_1, \ldots, x_{m+n}$. The system that generates the data can be described as: $y = f(x_1, x_2, \ldots, x_{m+n}) + \epsilon$, where ϵ is iid (independent and identically distributed) random error following $N(0, \sigma^2)$.

The single valued deterministic function of f with $m + n$ dimensional arguments portrays the joint predictive relationship of score y on target values for database and context attribute values $x_1, ..., x_n, ..., x_{m+n}$. In practical applications, a wide range of studies often encounter the problem of reconstructing an unknown function f from a finite set of discrete data. In the following, let's review the fruitful results achieved by the statistics community.

A Brief Review of Regression Models. Regression Analysis is one of the most widely used methods of investigating the statistical relationship between a set of predictor variables and a response variable. It is most often helpful when the predictor variables cannot be controlled, as when they are sampled in the observational study. The form of the model that is thought to relate the response variable to a set of predictor variables can be specified initially by the experts in the area of study based on their knowledge or their objective and/or subjective judgements. The hypothesized model can then be either confirmed or refuted by the analysis of the collected data. We need to select the form of the function $f(x_1, x_2, ..., x_r)$. This function can be classified into two categories: *linear* and *nonlinear*.

I. *Linear Regression Model*

Let $x_1, x_2, ..., x_r$ be r predictor variables or regressors thought to be related to a response variable Y. The linear regression model with a single response takes the form

$$Y = \beta_0 + \beta_1 x_1 + ... + \beta_r x_r + \epsilon$$

The term "linear' refers to the fact the models are linear in the unknown parameters $\beta_0, \beta_1, ... \beta_r$.

With n independent observations on Y and the associated regressor variable $x_{ij} (x_{ij}$ denotes the ith observation of variable x_j), the complete model is

$$Y_1 = \beta_0 + \beta_1 x_{11} + \beta_2 x_{12} + ... + \beta_r x_{1r} + \epsilon_1$$
$$Y_2 = \beta_0 + \beta_1 x_{21} + \beta_2 x_{22} + ... + \beta_r x_{2r} + \epsilon_2$$
$$\vdots \qquad \vdots \qquad \vdots \qquad \vdots$$
$$Y_n = \beta_0 + \beta_1 x_{n1} + \beta_2 x_{n2} + ... + \beta_r x_{nr} + \epsilon_n$$

where the error terms are assumed to have the following properties:
1. $E(\epsilon_j) = 0$;
2. $Var(\epsilon_j) = \sigma^2$;
3. $Cov(\epsilon_j, \epsilon_k) = 0, j \neq k$.
In matrix terms,

$$\mathbf{y} = \mathbf{X}\beta + \epsilon$$

II. *Polynomial Regression Model*

The model previously considered have specified a linear relationship between the predictor variables and the response variable. This restriction excludes many nonlinear mathematical forms. Polynomial regression model extends the simple linear regression to include higher order terms. The second-degree polynomial model in two variables is:

$$Y_i = \beta_0 + \beta_1 X_{i1} + \beta_2 X_{i2} + \beta_{11} X_{i1}^2 + \beta_{12} X_{i1} X_{i2} + \beta_{22} X_{i2}^2 + \epsilon_i$$

III. *Multivariate Adaptive Regression Splines*

Multivariate Adaptive Regression Splines (MARS) was first introduced by Friedman [11] to efficiently approximate the relationship between a dependent variable and a set of explanatory variables in a piece-wise regression. MARS is an adaptive procedure for regression, and is well suited for high dimensional problems(i.e., a large number of inputs), where the curse of dimensionality would probably create problems for other techniques.

MARS is a nonparametric regression procedure that makes no assumption as to how the dependent variables are related to predictors. Because the relationship between the context values, attributes and score value is unknown, MARS is particularly fit for approximating the scoring function f when parametric models are helpless.

MARS uses "hockey stick" basis functions of the form $(x - t)_+$ and $(t - x)_+$ which are the building blocks of the model. The "+" denotes positive part, so

$$(x - t)_+ = \begin{cases} x - t, & \text{if } x > t, \\ 0, & \text{otherwise} \end{cases}$$

$$(t - x)_+ = \begin{cases} t - x, & \text{if } x < t, \\ 0, & \text{otherwise} \end{cases}$$

Each function is piecewise linear, with the parameter t as the knot of the basis function. The pair of functions is called a *reflected pair*[12]. The set of basis functions is

$$\mathcal{F} = \{(X_j - t)_+, (t - X_j)_+\}_{\substack{t \in x_{1j}, x_{2j}, \ldots, x_{Nj} \\ j = 1, 2, \ldots, p.}}$$

Then, we use functions from the set $\mathcal{F}$ and their products to build the model. Thus the model takes the form

$$f(X) = \beta_0 + \sum_{m=1}^{M} \beta_m B_m(X),$$

where each $B_m(X)$ is a function in $\mathcal{F}$, or a product of several such functions. The algorithm consists of four steps:

Step 1: Start with the $B_0(X) = 1$.
Step 2: Search the space of basis functions for each predictor variable and split point. β_m for each B_m is estimated by the minimization of the residual sum-of-squares.
Step 3: Run Step 2 recursively until a model with some preset maximum number of terms is built.
Step 4: The basis functions which contribute least to the accuracy of the fit are removed.

4   Context Space Reduction

To learn an appropriate contextual ranking function, it is desirable to reduce context space first due to the following reasons. First, in many real situations, a context attribute is highly correlated to a certain database attribute. For example, users at the areas with very high ultra-violet radiation levels are usually more interested in high SPFs provided by sunscreen lotions. Second, a subset of context attributes is usually adequate for producing accurate prediction within a reasonable cost, as demonstrated by our experiment. Third, data collection usually leads to an information overload. Working with fewer context dimensions can not only increase computational efficiency but also make it easier to identify interesting patterns. Fourth, in complex situations, the relationships between context attributes and database attributes tend to be obscure, and it is not easy to manually select context attributes which affect the ranking of query results.

As a starting point, we make two assumptions related to contextual ranking functions in this study: 1) context attribute set and database attribute set have a *bijective* relationship, which is pre-specified by users; 2) The bijective association between a database attribute and a context attribute has a *monotonic* property.

Definition 4. *Context attribute set C and database attribute set A have a **bijective** relationship, if and only if every element $C_j \in C$ has a relationship with only one corresponding element $A_i \in A$, and vice verse. We use (A_i, C_j) to denote the bijective pair of database attribute A_i and context attribute C_j.*

With the *bijective relationship* assumption, the examination of context influence upon database query results can be translated into individually examining the effect of each context attribute upon its corresponding bijective database attribute.

Definition 5. *(A_i, C_j), where $(A_i \in A)$ and $(C_j \in C)$, is called a **monotonic bijective pair** if (A_i, C_j) is a bijective pair; and meanwhile for $\forall t, t' \in r(R)$, $\forall u, u' \in c(CS)$, $f(t.A_i, u.C_j) \leq f(t'.A_i, u'.C_j)$ subject to either of the following two conditions: 1) if $(t.A_i \leq t'.A_i)$, then $(u.C_j \leq u'.C_j)$; or 2) if $(u.C_j \leq u'.C_j)$, then $(t.A_i \leq t'.A_i)$.*

Referring to Table 1 and 2, the *product* attribute SPF and context attribute UVI is a monotonic bijective pair: the higher the value of context attribute UVI is, the higher the value of *product*'s SPF is preferred by users, leading to a higher rank value assigned to the corresponding *product* tuple.

Therefore, our context selection task degrades to search for a subset of bijective context and database attribute pairs according to a certain criterion. With each state in the search space specifying a subset of bijective pairs, the search space is $O(2^N)$ where N is the number of bijective context and database attribute pairs. When N is large, exhaustive search will be computationally prohibitive. Before presenting our context selection method, we first introduce some notions.

Definition 6. *Given a list of comparable values $L = \langle l_1, l_2, \ldots, l_x \rangle$, we define a **position list function** over L, $Posi(L) = L'$, which maps L to a new list L', whose elements are non-negative integers, indicating the ordering positions of the elements in L. We denote the position of element l_i of L by $posi(l_i, L)$. That is, $Posi(L) = \langle posi(l_1, L), posi(l_1, L), \ldots, posi(l_x, L) \rangle$.*

Example 3. *Given a list $L = \langle 100.0, 80.0, 200.5 \rangle$, $posi(100.0, L) = 2$, $posi(80.0, L) = 1$, and $posi(200.5, L) = 3$. Therefore, $Posi(L) = \langle 2, 1, 3 \rangle$.*

Definition 7. *Let $L = \langle l_1, l_2, \ldots, l_x \rangle$ and $L' = \langle l'_1, l'_2, \ldots, l'_x \rangle$ be two lists of the same length. A **minus** operator upon L and L' is defined as: $L - L' = \langle l_1 - l'_1, l_2 - l'_2, \ldots, l_x - l'_x \rangle$.*

Apparently, the binary $minus$ operator defined above is a closed operator.

Definition 8. *Given a set of contextually ranked database tuples of the form $[t, u, s]$ where $(t \in T) \wedge (u \in U) \wedge (s \in [0, 1])$, let $L_{A_i} = \langle t_1.A_i, t_2.A_i, \ldots, t_{|T|}.A_i \rangle$ be a list of database attribute A_i's values from $Dom(A_i)$; let $L_{C_j} = \langle u_1.C_j, u_2.C_j, \ldots, u_{|T|}.C_j \rangle$ be a list of context attribute C_j's values $Dom(C_j)$; and let $L_S = \langle s_1, s_2, \ldots, s_{|T|} \rangle$ be a list of ranked values given by users beforehand. We define the **impact factor** of context attribute C_j upon database attribute A_i, given $[T, U, S]$, as:*
$$\mathcal{I}mp(A_i, C_j)|_{[T,U,S]} = |r_s(Posi(L_{C_j}) - Posi(L_{A_i}), L_S)|.$$
Here, r_s is the well-known Spearman rank correlation coefficient [13] defined as: $r_s(L_1, L_2) = 1 - 6 \sum \frac{d^2}{N(N^2-1)}$, where d is the difference in statistical rank of corresponding variables in L_1 and L_2, and N is length of L_1.

Through the impact factor of a context attribute upon a database attribute, we can quantitatively measure their association strength in terms of whether a context attribute value coordinates well with the corresponding database attribute value. In other words, when their positions' difference within respective value lists is small, the ranking score is high. According to the impact factors of context attributes upon their bijective database attributes, we can identify those influential context and database attribute pairs whose impact factors are above a certain threshold σ, i.e., $\mathcal{I}mp(A_i, C_j)|_{[T,U,S]} \geq \sigma$. Algorithm 1 shows how to make context-database attribute pair selection in order to reduce context dimensionality.

Algorithm 1. The *ContextSpaceReduction Algorithm*

Input: a set of monotonic bijective context and database attribute pairs, a set of contextually ranked tuples of the form $[t, u, s]$ where $(t \in T) \wedge (u \in U) \wedge (s \in [0, 1])$;

Output: a selected subset of monotonic bijective context and database attribute pairs.

1: **for all** bijective pairs (A_i, C_j) **do**
2:     $L_{A_i} \leftarrow \langle t_1.A_i, t_2.A_i, \ldots, t_{|T|}.A_i \rangle$
3:     $L_{C_j} \leftarrow \langle u_1.C_j, u_2.C_j, \ldots, u_{|T|}.C_j \rangle$
4:     $L_S \leftarrow \langle s_1, s_2, \ldots, s_{|T|} \rangle$
5:     $L_{diff} \leftarrow Posi(L_{A_i}) - Posi(L_{C_j})$
6:     compute $r_s(L_{diff}, L_S)$
7:     **if** $|r_s(L_{diff}, L_S)| < \sigma$ **then**
8:         drop C_j
9:     **end if**
10: **end for**

Given a set of contextually ranked database tuples T under context, and let w be the total number of bijective pairs of context attributes and database attributes. The function *Posi* executes the sorting operation which takes $O(|T| * \log|T|)$ time complexity. Other operations within the inner loop take linear time. Therefore, the time complexity of the context selection algorithm is $O(w * |T| * \log|T|)$.

5   Evaluation

We did some experiments on both synthetic and real data to test the efficiency and effectiveness of our approach. All the tests were performed on a computer with Intel Core Duo CPU L7500 1.60GHz and 2G of Memory. The computer runs Windows Vista, and the algorithms were implemented in S language.

5.1   Experimental Data

Synthetic Data. Taking the running example in Section 1 for reference, assume we have two monotonic bijective pairs, (SFP, UVI) and $(Price, Income)$. The behavior of such monotonic bijective relationships implies that, the closer the values of v_{SPF} and $k * v_{UVI} - a$ (or v_{Price} and $l * v_{Income} - b$) are, the higher the contextual ranking function shall return. For simplicity, here, we use v_{SPF} to represent a value of the database attribute SFP related to a *product* tuple, and v_{UVI} to represent a value of the context attribute UVI under a certain context instance. k, l, a, b are constant values. We generate 3 synthetic data sets by following different forms of ranking functions.

Dataset 1. The contextual ranking function is set to $f(v_{SPF}, v_{UVI}) = -|v_{SPF} - k * v_{UVI} + a| * 0.935 + 100.311$ (where k=7.667, a=-15). The values of UVI and SFP are uniformly generated in the range of $[0, 15]$ and $[15, 130]$, respectively. In total, we generate 10k records in the form of $(v_{SPF}, v_{UVI}, f(v_{SPF}, v_{UVI}))$. Figure 1(a) demonstrates Dataset 1.

Dataset 2. The contextual ranking function is set to $f(v_{SPF}, v_{UVI}) = 5/|v_{SPF} - k * v_{UVI} + a|$ (where k=7.667, a=-15). The setting of UVI and SFP values is the same as in Dataset 1, with totally 10k records generated. Figure 1(b) shows uniformly sampled 200 records of Dataset 2.

Dataset 3. Besides the bijective pair (SFP, UVI), we incorporate another bijective pair $(Price, Income)$ into the contextual ranking function: $f(v_{SPF}, v_{Price}, v_{UVI}, v_{Income}) = 10/(|v_{SPF} - k * v_{UVI} + a| + |v_{Price} - l * v_{Income} + b|)$ (where $k = 7.667, l = 0.0095, a = -15, b = -10.5$). The values of v_{Price} and v_{Income} are also generated with a uniform distribution in the range of $[20, 200]$ and $[1000, 20000]$, respectively. The total number of records in the form of $(v_{SPF}, v_{Price}, v_{UVI}, v_{Income})$ varies from 100k to 500k.

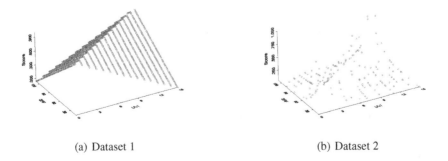

(a) Dataset 1   (b) Dataset 2

Fig. 1. Scatter plot of synthetic data

To test context space reduction, we meanwhile add one more context attribute value v_{Age} in the range of $[15, 90]$, and one more product attribute value $v_{ProductionDate}$ in the range of $[2004, 2008]$ to each record. We also added Gaussian random noises to the synthetic data generation and witness that the experimental results are not easily disturbed by the noises.

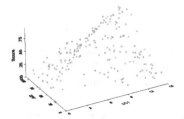

Fig. 2. Scatter plot of real data

Real Data. We invited 10 people to rank *product* tuples of different attribute values under different context instances, which are organized in the following 4 groups:
1) (v_{SPF}, v_{UVI}); 2) (v_{Price}, v_{Income}); 3) $(v_{ProductionDate}, v_{Age})$;
4) $(v_{SPF}, v_{Price}, v_{ProductionDate}, v_{UVI}, v_{Income}, v_{Age})$. Each group has 20 (record) * 10 (people) = 200 records for learning and testing the contextual ranking function. Figure 2 plots our surveyed data of $(v_{UVI}, v_{SPF}, score)$ in 3-dimensional space with UVI as x-axis, SPF as y-axis, and *score* as z-axis.

For both synthetic and real data sets, we take 80% of each to learn the contextual ranking function, and 20% for testing purpose.

5.2   Basic Experiments

Three regression models described in Section 3, i.e., the Generalized Linear Model (GLM), polynomial regression, and Multivariate Adaptive Regression Splines (MARS), are exploited to learn the contextual ranking function. We examine how good the regression models fit the data by means of the *deviance residual* measurement, recommended by McCullagh and Nelder [19]. The deviance value is written in the form of $\sum_{i=1}^{n} d_i = D(\beta)$, where $d_{i,r} = [\text{sgn}(y_i - \hat{\mu}_i)] \cdot \sqrt{d_i}, i = 1, 2, ..., n$ is called deviance residual [20].

Results on both synthetic data and real data - Group 1 $(v_{SPF}, v_{UVI}, score)$, as shown in Figure 3, demonstrate the feasibility of regression models for learning contextual ranking functions, where MARS and polynomial regression are more robust than GLM under different datasets.

5.3   Performance of Context Selection

We ran the experiments on synthetic Dataset 3 to investigate how much we can gain by context space reduction. The original Dataset 3 contains three context and database

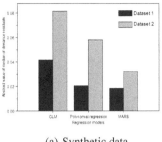
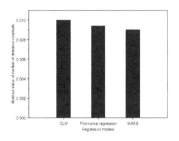

(a) Synthetic data   (b) Real data

Fig. 3. Absolute value of median of deviance residuals on synthetic and real data

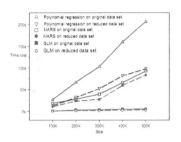
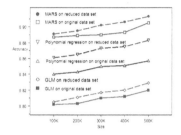

(a) Time cost on synthetic data   (b) Accuracy on synthetic data

Fig. 4. Performance comparisons with/without context selection on synthetic data

attribute pairs. The context selection algorithm will eliminate the pair ($ProductionDate$, Age). In Figure 4(a), blue, red, and green solid lines stand for the performances of polynomial regression, MARS, and GLM, respectively before context selection. Their dashed counterparts stand for the performances after context space reduction. Experimental results justify that our context selection algorithm can greatly increase computational efficiency, especially for MARS and polynomial regression which usually need to take longer time than GLM.

Besides efficiency, we also study the effectiveness of context selection(where $\sigma = 0.85$). We call the output ranking result an *inversion ranking* if t is preferred to t' in testing data but t' is preferred to t in the predicting output. With it, we define an *Accuracy* metric as one minus the proportion of the number of inversion rankings to the maximum possible number of inversions rankings.

From the results presented in Figure 4(b), it is interesting to see that context selection can not only increase execution efficiency, but also improve the ranking effectiveness compared to the non-context selection strategy.

We test the context selection strategy on the real data - Group 4. Due to the impact factor of context attribute Age upon $ProductionDate$ is less that σ, which is set to 0.85, we eliminate this pair. Figure 5 shows almost no loss in accuracy after we reduce the context space.

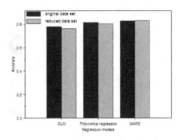

Fig. 5. Performance comparisons with/without context selection on real data

6   Conclusion

Studying the impact of context attributes upon certain database attributes is fundamental and crucial to context-aware database querying. In this paper, we exploit the use of regression models developed by the statistics community to contextually rank database query results. We propose to reduce the context search space so as to increase efficiency and meanwhile enhance our understanding of the inherent relationship between contexts and attributes. Unlike traditional principal component analysis methods to reduce dimensionality, our algorithm is based on rank correlation coefficients to eliminate irrelevant contexts and avoid transformation which could make context information less interpretable. We evaluate the performance of the approach and demonstrate that our algorithm increases the efficiency and effectiveness of regression models with dimension reduction. In the future, we plan to investigate different dimensionality reduction techniques on hierarchical contexts, and integrate qualitative and quantitative approaches together to deal with context-aware database querying.

References

1. Agrawal, R., Rantzau, R., Terzi, E.: Context-sensitive Ranking. In: ACM SIGMOD 2006, Chicago, Illinois, USA (2006)
2. Agrawal, R., Wimmers, E.: A Framework for Expressing and Combining Preferences. In: ACM SIGMOD 2000, Dallas, Texas, USA (2000)
3. Bellman, R.: Adaptive Control Processes: A Tour Guide. Princeton University Press, Princeton (1961)
4. Bunningen, A., Feng, L., Apers, P.: A context-aware query preference model for Ambient Intelligence. In: Bressan, S., Küng, J., Wagner, R. (eds.) DEXA 2006. LNCS, vol. 4080. Springer, Heidelberg (2006)
5. Bunningen, A., Fokkinga, M., Apers, P., Feng, L.: Ranking Query Results using Context-Aware Preferences. In: First Intl. Workshop on Ranking in Databases (In Conjunction with ICDE 2007), Istanbul, Turkey, April 16 (2007)
6. Cai, D., He, X., Han, J.: Spectral Regression: A Unified Subspace Learning Framework for Content-Based Image Retrieval. In: ACM Multimedia 2007, Augsburg, Germany (September 2007)

7. Chen, K., Zhang, Y., Zheng, Z., Zha, H., Sun, G.: Adapting Ranking Functions to User Preference. In: Second Intl. Workshop on Ranking in Databases(DBRank 2008), Cancun, Mexico (2008)
8. Chomicki, J.: Preference formulas in relational queries. ACM Trans. Database Syst. 28(4), 427–466 (2003)
9. Cunningham, P.: Dimension Reduction. Technical Report UCD-CSI-2007-7 University College Dublin (2007)
10. Feng, L., Apers, P., Jonker, W.: Towards Context-Aware Data Management for Ambient Intelligence. In: Galindo, F., Takizawa, M., Traunmüller, R. (eds.) DEXA 2004. LNCS, vol. 3180. Springer, Heidelberg (2004)
11. Friedman, J.: Multivariate Adaptive Regression Splines. The Annual of Statistics 19(1), 1–67 (1991)
12. Hastie, T., Tibshirani, R., Friedman, J.: The Elements of Statistical Learning. Springer, Heidelberg (2001)
13. Higgins, J.: Introduction to Modern Nonparametric Statistics. Duxbury Press (2003)
14. Joachims, T.: Optimizing Search Engines using Clickthrough Data. In: SIGKDD 2002, Edmonton, Alberta, Canada (2002)
15. Kießling, W.: Foundations of Preferences in Database Systems. In: VLDB, pp. 311-322, Hong Kong, China (2002)
16. Koutrika, G., Ioannidis, Y.: Personalized Queries under a Generalized Preference Model. In: Proc. of 21st Intl. Conf. On Data Engineering (ICDE), Tokyo, Japan, April 5-8 2005, pp. 841–852 (2005)
17. Lattin, J., Carroll, J., Green, P.: Analyzing Multivariate Data. Duxbury (2003)
18. Liu, H., Motoda, H.: Feature Selection for Knowledge Discovery and Data Mining. Kluwer Academic Publishers, Dordrecht (1998)
19. McCullagh, P., Nelder, J.: Generalized linear Models, 2nd edn. Chapman & Hall/CRC, London (1989)
20. Myers, R., Montgomery, D., Vining, G.: Generalized Linear Models with Applications in Engineering and the Sciences. John Wiley & Sons, Chichester (2002)
21. Stefanidis, K., Pitoura, E.: Fast Contextual Preferences Scoring of Database. In: EDBT 2008, Nantes, France (2008)
22. Stefanidis, K., Pitoura, E., Vassiliadis, P.: On Supporting Context-Aware Preferences in Relational Database Systems. In: First Intl Workshop on Managing Context Information in Mobile and Pervasive Environments(MCMP 2005), Ayia Napa, Cyprus (2005)
23. Stefanidis, K., Pitoura, E., Vassiliadis, P.: Adding Context to Preferences. In: ICDE, Istanbul, Turkey (2007)

On the Evaluation of Quality of Context*

Atif Manzoor, Hong-Linh Truong, and Schahram Dustdar

Distributed Systems Group, Vienna University of Technology
{manzoor,truong,dustdar}@infosys.tuwien.ac.at

Abstract. High quality context information plays a vital role in adapting a system to the rapidly changing situations. However, the diversity of the sources of context information and the characteristics of the computing devices strongly impact the quality of context information in pervasive computing environments. Quality of Context parameters can be used to characterize the quality of context information from different aspects. In this paper, we quantify the Quality of Context parameters to present them in a suitable form for use with the applications in pervasive environments. We also present a mechanism to tailor the Quality of Context parameters according to the needs of an application and then evaluate these parameters. Enrichment of context information with Quality of Context parameters enhances the capabilities of context-aware systems to effectively use the context information to adapt to the emerging situations in pervasive computing environments.

1 Introduction

Designing an adaptive system in pervasive computing environments faces a stern challenge because of the limitations of the quality of context information [2]. Context information has an innate characteristic of imperfection [4] and its quality is highly influenced by the way it is acquired in pervasive computing environments [28]. Diverse sources of context information, ranging from physical and logical sensors to user interfaces and applications on mobile devices, also affect the quality of the context information and context information can be imperfect [13].

Many research efforts have been done to provide and use context information to achieve the vision of pervasive computing environments since it has been presented in [29]. However, as indicated in [1], existing systems rarely give any consideration to the quality of context information. Previous studies show that the applications using context information either make wrong assumptions about its quality [5, 7, 13] or perform extra effort to deal with the uncertainty of context information [20].

As discussed in [14], "context information is (a) incorrect if it fails to reflect the true state of the world it models, (b) inconsistent, e.g., by containing contradictory information, or (c) incomplete if some aspects of context are unknown". Therefore, possessing the knowledge about the quality of the context information plays an important role in using that information effectively and achieving the capability of context awareness.

---

* This research is partially supported by the European Union through the FP6-2005-IST-5-034749 project WORKPAD.

D. Roggen et al. (Eds.): EuroSSC 2008, LNCS 5279, pp. 140–153, 2008.

Subsequently, research efforts have been undertaken to model the quality of context information [12, 22]. The term "Quality of Context(QoC)" has been coined in [3] and is defined as "any information that describes the quality of information that is used as context information". QoC parameters have also been listed as the probability of correctness, trust-worthiness, resolution, and up-to-dateness [3]. However, few works [16, 23] have discussed the evaluation of QoC parameters. There has not been any work which presents and evaluates the QoC parameters as the worth of context information for an application and provides the context information enriched with these QoC parameters.

In this paper we evaluate the QoC and relate QoC to the worth of context information for an application. We classify QoC into QoC sources and parameters. QoC sources are the information about the sources that collect context information, the environments where that context information is collected, and the entities about which the context information is collected. These QoC sources are used to evaluate QoC parameters that indicate the worth of context information for an application. Context information enriched with QoC parameters is provided to the applications to make them aware of the quality of context information they use. We also discuss our context information model that represents the context information enriched with QoC parameters. We have implemented our approach to provide context information enriched with QoC parameters to support adaptiveness in disaster response activities.

The rest of paper is organized as follows: Section 2 illustrates a scenario and discusses the motivation for our work. Section 3 gives an overview of related work and compares them with our approach. Section 4 presents a classification of QoC sources and QoC parameters. Our approach to evaluate the QoC parameters is presented in Section 5. We discuss our context information model enriched with QoC parameters in Section 6. Our prototype implementation is explained in Section 7. Finally, we conclude the paper and discuss future work in Section 8.

2 Motivation and Scenario

In August 2002, widespread persistent rain led to the catastrophic floods in many parts of Central Europe. There were extreme rainfall events in Austria on numerous rivers north of the Central Alps starting from the west. The northern Federal Provinces of Upper and Lower Austria as well as the Federal Province of Salzburg were particularly affected. This event brought rainfall of extraordinary extent and flood recurrence intervals from several years to more than 100 years. Loses of human life and livestock, damages on the infrastructure, buildings, public and private properties rose the public awareness and the demand for improvement of future flood mitigation measures, innovative alert systems and new technological solutions needed to improve the rescue activities and the analysis of the damage caused by the floods.

This scenario is illustrated in [11] and is supported in the EU project WORKPAD [25]. In this project, we aim at facilitating people, performing rescue work in such situations, by providing context information to them. Field workers, participating in the rescue activities and equipped with mobile devices, share context information among themselves and send it to the back end by using context information management services as described in our ESCAPE framework [27]. But due to the unawareness about the quality of information, people face problems in using this information. For example, in afore mentioned scenario, a geographical information specialist makes an analysis of the damage caused by the flood and provides this information to the organizations participating in the rescue activities. He receives context information from the field workers and combines it with the data stored in a geographical information system (GIS) to update the current flood situation. As there are usually more than one context source providing information of the same entity in the field, he has to make analysis of all information to be aware of the quality of that information. The photos taken all look the same and he has often not been able to select one of them and relate it to the digital map stored in the GIS. These problems not only cost him time and effort, but also affect the quality of his work. Context information, enriched with QoC parameters such as up-to-dateness, trust-worthiness, completeness, and significance, allows him to know the quality of context information without looking at the contents of context information, and thus substantially improving his work.

3   Related Work

Quality of context information, gathered in pervasive computing environments, has been considered unsatisfactory since the start of research in context-aware systems [7, 14]. It has also been considered a challenge to deal with this shortcoming of the quality of context information in the development of context-aware systems [8, 12]. But only few works [14, 8] have used metadata to indicate characteristics of context information and advantages of modeling metadata with context information, e.g., [15]. Subsequently, problems associated with this imperfection of context information and its causes have been explored in more detail [13]. Here, we discuss the works that defined the term Quality of Context, identified the parameters to indicate the Quality of Context, and evaluated some of them.

3.1   Quality of Context

Quality of Context (QoC) was first defined in [3] as "any information that describes the quality of information that is used as context information". Later on, QoC has also been defined in [17] as "any inherent information that describes context information and can be used to determine the worth of information for a specific application". Precision, probability of correctness, trust-worthiness, resolution, and up-to-dateness has been identified as important QoC parameters in [3]. This list has been extended to include accuracy, completeness, representation consistency, and access security selected on the basis of user's concern in the quality of context information [16]. The characteristics of the sensor, the situation of the specific measurement, the values expressed by the context information object itself, and the granularity of the representation format have also been identified as the sources to determine QoC [17].

However, these works have not provided the QoC parameters in a form that can show the worth of context information for an application and to allow those parameters to be used by an application. They have also not made any distinction between the QoC indicators which can be used to calculate high level QoC parameters for application usage.

3.2 QoC Evaluation

Though QoC parameters have been indicated and context information models have been designed to accommodate these parameters with context information [12, 21, 24, 30], few works have tried to evaluate these parameters. In [22], incompleteness, inconsistency and variation in precision of context information have been identified as sources of imperfection in context information. A three-layered model, for user context management, has been discussed to handle this problem of imperfect context information. However, only the age of context information has been used to measure the confidence in context information that is not enough to indicate the quality of context information in pervasive computing environments.

An analysis of QoC indicators, such as precision, freshness, spatial resolution, temporal resolution, and probability of correctness and different options to quantify these parameters, has been done in [23]. Later, this work has used these QoC parameters to enforce the user privacy policy in a health tele-monitoring scenario. But no mechanism has been provided to evaluate these parameters for context information and provide these parameters to the applications. As compared to their work, we have also related QoC parameters to the worth of context information for an application.

A relationship between dimensions of information quality and QoC parameters have been discussed in [16]. Accuracy, completeness, representation consistency, access security, and up-to-dateness have been selected as quality dimensions of context information. This work used a statistical estimation method to calculate the accuracy of sensor data in smart homes. However, their method to measure accuracy is only appropriate in those cases where sensors get continuous data around some average value, e.g., via temperature sensors. Completeness is also measured as the ratio of available attributes to total number of attributes for a specific context object. This work has also not provided the evaluation of enough QoC parameters to be useful with the applications in pervasive computing environments.

4 QoC Classification

As we aim at relating QoC to the worth of context information for an application, we classify QoC into QoC sources and QoC parameters. QoC sources are the quantities which are sensed from the environment or profiled as the configuration of a system. But as these values are not appropriate for use with an application, they are transformed to higher level QoC parameters that are in more suitable form for computational use. Context information is then annotated with these QoC parameters and is provided to users or applications.

Let $\mathcal{O}$ be a context object that has context information about a real world entity $\mathcal{E}$. The context object $\mathcal{O}$ is collected by a sensor $\mathcal{S}$. Sensor $\mathcal{S}$ can be a physical sensor or a logical sensor, e.g., user interfaces and applications on a portable device.

4.1 QoC Sources

QoC sources describe the information about the source of context information, environment in which context information is gathered and a context information object itself.

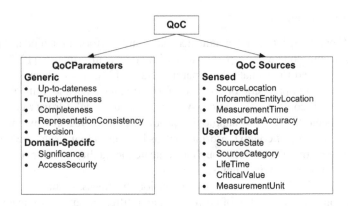

Fig. 1. Classification of QoC into sources and parameters

They include information that have been sensed from the environment, derived from existing information, or profiled at the source of context information. These values will be used to determine the QoC parameters. Figure 1 shows the QoC sources which are collected along with context object. Here we will give a short description of each of them.

SourceLocation is the geographical location of the source that collects the context object and *InformationEntityLocation* is the geographical location of the entity that is represented by that context object. *SourceLocation* along with *InformationEntityLocation* will be representing the space resolution. They help to decide about the trust-worthiness on the context object. For example, if we have more than one context object, representing the same entity in the environment, the context object reported by the source that is nearest the entity will get maximum value of trust-worthiness, provided that all the sources are collecting the information with the same accuracy.

MeasurementTime is the time at which context information is measured. *SensorDataAccuracy* is the accuracy with which a sensor can collect a context object. *SourceState* indicates whether the source of information is dynamic or static. For example, sensors measuring temperature are fixed at different places in a city. These sources have static value for the *SourceState*. While sensors embedded in a portable device carried by a human have dynamic value for the *SourceState*, e.g., GPS sensors embedded in mobile phones. *SourceCategory* indicates category that a source of context information can have in pervasive computing environments. Possible values for the *SourceCategory* can be sensed, profiled, derived, and static [13]. *LifeTime* is the period of time after which context information becomes obsolete and it is necessary to take its value again. For example, the location of fast moving vehicle may has less value regarding lifetime as compared to the location of a walking person.

CriticalValue of context information will indicate that this information is crucial in a specific scenario. This concept particularly affects quality of context information in scenarios where it will be used in tasks involving saving the human lives. For example, in our afore mentioned scenario, context object having information about the people caught in the low lying area of the city will by of high critical value. *MeasurementUnit* is used to describe precision of context information. The location of an entity measured

Table 1. Association between the QoC parameter and sources

QoC parameter	QoC sources used for evaluation
Up-to-dateness	MeasurementTime, CurrentTime
trust-worthiness	SourceLocation, InformationEntityLocation, SensorDataAccuracy
completeness	Ratio of number of attributes filled to total number of attributes
significance	CriticalValue

with the accuracy of ten meters will have less precision as compared to a measurement up to the accuracy of one meter.

4.2 QoC Parameters

QoC parameters are derived from QoC sources and are represented in a form that is suitable for use by an application. Figure 1 shows QoC parameters divided into generic and domain specific parameters. Generic QoC parameters are those parameters which are required by most applications, such as precision, trust, validity, representation consistency, and completeness [3]. Domain specific QoC parameters are those parameters that are important for some specific application domains, such as significance and access security. Generic QoC parameters have been described in different works [3, 17, 23]. Significance of context information is important in life threatening situations such as in emergency response. It signals unusual events, in the environment, which need immediate attention. For example, in a flood response scenario if water level raises more than threshold value, this information will be of high significance and will need immediate response. In the next section we will discuss how can we quantify and evaluate some of the QoC parameters.

5 QoC Evaluation

In this section we discuss our approach and system that evaluates QoC parameters. The first step to represent QoC parameters in such a form that they can easily be used by an application is to quantify them. Quality is a relative term and is measured against some standards so it is appropriate to measure QoC parameters as a decimal which can have value in the range [0..1]. Maximum value 1 means that QoC parameter is in complete compliance to the given requirements while minimum value 0 means total nonconformity to the requirements. Table 1 shows the QoC parameters which are evaluated in this section and QoC sources which are used for their evaluation. We have selected those QoC parameters for close examination which are more expressive and have substantial impact on our illustrated scenario.

5.1 Up-To-Dateness

This quality measure indicates the degree of rationalism to use a context object for a specific application at a given time. We take into account the age of context object and the lifetime of that context object to calculate the value of up-to-dateness. The age of the context object $\mathcal{O}$ is calculated by taking the difference between the current time, t_{curr}, and the measurement time of that context object $\mathcal{O}$, $t_{meas}(\mathcal{O})$, as shown by Equation 1.

$$Age(\mathcal{O}) = t_{curr} - t_{meas}(\mathcal{O}) \qquad (1)$$

Up-to-dateness of the context object $\mathcal{O}$, $\mathcal{U}(\mathcal{O})$, is calculated by Equation 2.

$$\mathcal{U}(\mathcal{O}) = \begin{cases} 1 - \frac{Age(\mathcal{O})}{Lifetime(\mathcal{O})} & : \quad if \ Age(\mathcal{O}) < Lifetime(\mathcal{O}) \\ 0 & : \quad otherwise \end{cases} \qquad (2)$$

The value of up-to-dateness and hence the validity of context object $\mathcal{O}$ decrease as the age of that context object increases. Therefore, the geographical information specialist in our scenario can look at the value of this QoC parameter to be sure of the validity of information contained by the context object. This parameter can help him to more confidently combine the information contained in this context object with the existing information in GIS to provide the current situation of flood in the city to teams participating in rescue work. Context objects, having low value of the up-to-dateness, may have misleading or wrong information and can be ignored. If we have static information, e.g., rescue worker's profile information saved at the back end, we can set the lifetime of that information infinite so that its age will not affect the value of up-to-dateness and it will always be maximum.

5.2 Trust-Worthiness

This quality measure indicates the belief that we have in the correctness of information in a context object. Trust-worthiness of a context object is highly affected by the space resolution, i.e., the distance between the sensor and the entity. The farther the sensor from the entity the more will be the doubt in the correctness of information presented by that context object. Along with the space resolution, accuracy, with which the sensor collects a context object, also affects the trust-worthiness of that context object. Let the accuracy of the sensor data be δ. The trust-worthiness, $\mathcal{T}(\mathcal{O})$, of context object $\mathcal{O}$ is defined by Equation 3.

$$\mathcal{T}(\mathcal{O}) = \begin{cases} (1 - \frac{d(\mathcal{S},\mathcal{E})}{d_{max}}) * \delta & : \quad if \ d(\mathcal{S},\mathcal{E}) < d_{max} \\ 0 & : \quad otherwise \end{cases} \qquad (3)$$

where $d(\mathcal{S},\mathcal{E})$ is the distance between the sensor and the entity. d_{max} is the maximum distance for which we can trust on the observation of this sensor. Every type of sensor will have different value for d_{max}. For example d_{max} value for a satellite capturing the photos from the space will be a lot more than the camera held by a work to take photos in the field. Accuracy of a sensor, δ, is measured on the basis of a statistical estimation method presented in [16].

Trust-worthiness is useful in situations when we have more than one context object representing the same entity. There will be more confidence in the context object collected by the sensor that has a higher value of trust-worthiness. For instance, in scenario discussed in Section 2, photos will be preferred from a worker that have higher value of trust-worthiness. It means that the worker is nearest the entity, represented by the photo, and he has device capability to capture the photo with more accuracy.

5.3 Completeness

This quality measure indicates the quantity of information that is provided by a context object. In [16] completeness has been computed as the ratio of the number of attributes available to the total number of attributes. We have enhanced this concept by using the weights for different attributes, as all attributes of a context object do not have the same importance. Hence, we define the completeness of a context object as the ratio of the sum of the weights of available attributes of a context object to the sum of the weights of all the attributes of that context object. Completeness, $\mathcal{C}(\mathcal{O})$, of context object $\mathcal{O}$ is evaluated by Equation 4.

$$\mathcal{C}(\mathcal{O}) = \frac{\sum_{j=0}^{m} w_j(\mathcal{O})}{\sum_{i=0}^{n} w_i(\mathcal{O})} \tag{4}$$

where m is the number of the attributes of context object $\mathcal{O}$ that have been assigned a value and $w_j(\mathcal{O})$ represents the weight of the jth attribute of $\mathcal{O}$ that has been assigned a value. Similarly, n is the total number of the attributes of context object $\mathcal{O}$ and $w_i(\mathcal{O})$ represents the weight of the ith attribute of $\mathcal{O}$. The value of completeness will be maximum, i.e., 1 if $n = m$. It means that all the attributes of context object $\mathcal{O}$ have been assigned a value.

5.4 Significance

This quality measure indicates the worth or the preciousness of context information in a specific situation. Its value is of particular importance in scenarios that involve life threatening situations for humans. For example, in the flood scenario described in Section 2, significance of the context object will increase if it contains information which needs immediate response or attention, e.g., collapse of a building in a low lying area of the city where the water level is high. A context object having this information gets a higher value of significance so that it will get immediate attention. Significance of context object $\mathcal{O}$, $\mathcal{S}(\mathcal{O})$, is evaluated by Equation 5.

$$S(\mathcal{O}) = \frac{CV(\mathcal{O})}{CV_{max}(\mathcal{O})} \tag{5}$$

where $CV(\mathcal{O})$ is the critical value of the context object $\mathcal{O}$. This information will be gathered from a situation configuration file. This configuration file will have the information about the critical values of each type of concept in the context model for a specific situation. $CV_{max}(\mathcal{O})$ is the maximum critical value that can be assigned to a context object of the type that is represented by $\mathcal{O}$.

5.5 Algorithm for the Evaluation of QoC Parameters

Algorithm 1 has been defined to evaluate the QoC parameters. This algorithm takes a context object $\mathcal{O}$ as input and requires that the QoC source information is attached to the context object. It calculates the values of QoC parameters and returns the context object after annotating it with these parameters.

Algorithm 1. Evaluation of QoC parameters

INPUT: ContextObject $\mathcal{O}$ with QoC sources
OUTPUT: ContextObject $\mathcal{O}$ with QoC parameters

1: $qocParameters \longleftarrow null$
2: $qocSources \leftarrow get\ QoCSources\ from\ \mathcal{O}$
3: **if** $qocSources \neq null$ **then**
4:     $t_{meas}(\mathcal{O}) \leftarrow get\ MeasurementTime\ of\ \mathcal{O}$
5:     $age \leftarrow t_{curr} - t_{meas}(\mathcal{O})$
6:     $lifetime \leftarrow get\ LifeTime\ of\ \mathcal{O}$
7:     **if** $age < lifetime$ **then**
8:         $up - to - dateness \leftarrow 1 - (age/lifetime)$
9:     **end if**
10:     {set value of validity in qocParameters}
11:     $qocParameters.Up - to - dateness \leftarrow up - to - dateness$
12:     $loc_s \leftarrow get\ SourceLocation\ from\ \mathcal{O}$
13:     $loc_e \leftarrow get\ InformationEntityLocaiton\ from\ \mathcal{O}$
14:     $distance \leftarrow distance\ between\ loc_s\ and\ loc_e$
15:     $accuracy \leftarrow get\ SensorDataAccuracy\ from\ \mathcal{O}$
16:     **if** $distance < distance_{max}$ **then**
17:         $trust - worthiness \leftarrow (1 - distance/distance_{max}) * acc$
18:     **end if**
19:     {set value of trust-worthiness in qocParameters}
20:     $qocParameter.Trust - worthiness \leftarrow trust - worthiness$
21:     **for** each attribute in $\mathcal{O}$ **do**
22:         **if** $value\ for\ attribute\ is\ available$ **then**
23:             {Sum of the weights of attributes which have the value}
24:             $w_a \leftarrow w_a + weight\ assigned\ to\ attribute$
25:         **end if**
26:         {Sum of the weights of all the attributes}
27:         $w_c \leftarrow w_c + weight\ assigned\ to\ atttribute$
28:     **end for**
29:     $completeness \leftarrow w_a/w_c$
30:     {set value of completeness as qocParameters}
31:     $qocParameter.Completeness \leftarrow completeness$
32:     $significance \leftarrow critical\ value\ of\ \mathcal{O}/maximum\ critical\ value\ that\ \mathcal{O}\ can\ have$
33:     {set significance as value of qocParameters}
34:     $qocParameters.Significance \leftarrow significance$
35:     {Annotation of ContextObject $\mathcal{O}$ with qocParameters}
36:     $\mathcal{O}.QoCParameters \leftarrow qocParameters$
37: **end if**
38: $RETURN\ \mathcal{O}$

At Line 1 the algorithm extracts QoC source information that is attached to context object $\mathcal{O}$. If QoC sources information is not provided then it returns the context object without the values of QoC parameters. At Line 1 it gets the measurement time of context object $\mathcal{O}$ that is used to calculate the age of the context object at Line 1. Validity is calculated at Line 1 and is added to QoC parameters at Line 1.

The algorithm gets the location of the sensor that gathers the context object at Line 1 and gets the location of the entity that is represented by the context object at Line 1. The distance between the sensor and the entity is calculated at Line 1 and is used to calculate the trust-worthiness of context object $\mathcal{O}$ at Line 1. Trust-worthiness is assigned to QoC parameters at Line 1. At Line 1 it calculates the sum of the weights of attributes that have been assigned a value while the sum of the weights of all attributes is calculated at Line 1. Completeness is calculated at Line 1 and assigned to QoC parameters at Line 1. Finally the significance of context object $\mathcal{O}$ is calculated at Line 1 and added to QoC parameters at Line 1. Context object $\mathcal{O}$ is enriched with QoC parameters at Line 1 and returned at Line 1.

6   Enrichment of Context Information Model with QoC

Developing a model to represent context information is of foremost importance in a context-aware system. Without a suitable model it is impossible to provide context information. The main considerations to model context information are: What are the system requirements? What is the purpose of collecting information? How is the information going to be used? Who will use this information? Another factor in the provisioning of context information is to deliver the user with only necessary information considering his current situation.

```
<xs:complexType name="QoC-Source">
    <xs:sequence>
        <xs:element name="InformationEntityLocation" type="ci:GeoLocationType"/>
        <xs:element name="Accuracy" type="ci:QoC-Decimal"/>
        <xs:element name="SourceLocation" type="ci:GeoLocationType"/>
    </xs:sequence>
    <xs:attribute name="SourceState" type="ci:SourceStateType" />
    <xs:attribute name="SourceCategory" type="ci:SourceCategoryType"/>
    <xs:attribute name="LifeTime" type="xs:duration"/>
    <xs:attribute name="CriticalValue" type="ci:CV_Integer"/>
    <xs:attribute name="MeasurementUnit" type="xs:string"/>
    <xs:attribute name="MeasurementTime" type="xs:dateTime"/>
</xs:complexType>
```

Fig. 2. (Simplified) XML schema representation of QoC source

To achieve these requirements, our model to provide context information for supporting disaster response is based on W4H [6] and granular context model [9]. The main concepts in this model are the *SupportWorkers* that perform rescue work on disaster site, *Devices* that are used by the *SupportWorkers* for communication and collection of context information, *Activities* of the *SupportWorkers*, and the *Site* of disaster response. Each concept is represented by a context object and can contain other context objects. For example, the context object providing information about a *SupportWorker* includes the context object providing information about the *Device* used by that *Support-Worker*. Each context object also include *QoCSources* information that are described in Section 4 and *QoCParameters* that are evaluated in Section 5.

```
<xs:complexType name ="QoCParametersType">
    <xs:sequence>
        <xs:element name="Up-to-dateness" type="ci:QoC-Decimal"/>
        <xs:element name="Trust-worthiness" type="ci:QoC-Decimal"/>
        <xs:element name="Completeness" type="ci:QoC-Decimal"/>
        <xs:element name="Significance" type="ci:QoC-Decimal"/>
    </xs:sequence>
</xs:complexType>
<xs:simpleType name="QoC-Decimal">
    <xs:restriction base="xs:decimal">
        <xs:minInclusive value ="0.0"/>
        <xs:maxExclusive value ="1.0"/>
    </xs:restriction>
</xs:simpleType>
```

Fig. 3. (Simplified) XML schema representation of QoC-parameters

Figure 2 shows the XML representation of the QoC sources that will be attached to every context object and Figure 3 shows the XML representation of the QoC-parameters. Every QoC-parameter is of type QoC-Decimal which is defined as the restricted type over decimal to have the value in range [0.0, 1.0].

7   Implementation

The components of our system that evaluate the QoC parameters (QoC Evaluator), annotate the context information (QoC Annotator), and provide QoC enriched context information (QoC Provider) to the applications using it are shown in Figure 4. We have implemented this system by using our Vimoware toolkit [26] that supports the development of Web services in ad-hoc networks of mobile devices. Sensors, collecting the information from the environment, are implemented as Context Information Management Service (CIMS). These sensors include the physical and logical sensors such as GPS sensor, battery sensor, memory sensor, network sensor, and sensors embedded in user interfaces and applications on the mobile devices.

CIMS publishes the type of context object, as defined by context information model, by using the *Context Publisher* component. Applications can query and subscribe for the context information using *Context Query and Subscriber*. Context information is provided to those applications by the *Context Provider*. *QoC Evaluator* evaluates the QoC parameters from the information provided in the context object about the QoC sources as described in Section 5. Sensors will be configured from a local configuration file which will have the information about *SourceState*, *SourceType* and components to evaluate the QoC parameters will be configured from a situation configuration file. This situation configuration file will have the information about *LifeTime* and *CriticalValue* of each type of context information concept in context information model for a specific scenario. Our mechanism to evaluate the QoC parameters can be configured according to the requirements of an application or situation by this configuration file. Finally, *Context Annotator* enriches the context information with already evaluated QoC parameters that is provided to subscribed applications by *QoC Provider*.

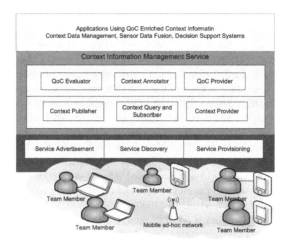

Fig. 4. Components for provisioning of QoC-parameters annotated context information

Sensors, providing information about the machine capabilities, have been implemented using the Intel Mobile SDK [10]. MXQuery [19] and KSOAP2 [18], having a low memory foot print to be able to run on mobile devices, have been used for processing XML data.

8   Conclusion and Future Work

In this work we have discussed the QoC parameters as the worth of context information for a specific application. Selected QoC parameters have also been evaluated and provided to the subscribed applications along with context information. Subsequently, enrichment of context information with Quality of Context parameters by our implemented algorithm enhances the perception of an application about the context information and enables the system to use this context information is an improved way.

For our next steps, we plan to extend this work with the evaluation of the remaining QoC parameters and perform the experiments to evaluate the effectiveness of these parameters. We will also broaden the scope of QoC parameters by discussing the role of reputation and historical information about the source of information and probabilistic nature of the context information in their evaluation. We will examine how the applications using these QoC parameters mention their weights according to their preference, how these parameters depend on each other, and how can we reach a tradeoff between their values. We plan to use QoC parameters in context data management and context information fusion. We will also enhance the quality by combining the context information and QoC parameters represented by more than one context object. We will use this high quality context information to provide robustness in context-aware systems.

References

1. Baldauf, M., Dustdar, S., Rosenberg, F.: A survey on context-aware systems. Int. J. Ad Hoc Ubiquitous Comput. 2(4), 263–277 (2007)
2. Bu, Y., Gu, T., Tao, X., Li, J., Chen, S., Lu, J.: Managing quality of context in pervasive computing. In: QSIC 2006: Proceedings of the Sixth International Conference on Quality Software, Washington, DC, USA, pp. 193–200. IEEE Computer Society, Los Alamitos (2006)
3. Buchholz, T., Kpper, A., Schiffers, M.: Quality of context: What it is and why we need it. In: Proceedings of the 10th International Workshop of the HP OpenView University Association(HPOVUA), Hewlet-Packard OpenView University Association (2003)
4. Castro, P., Muntz, R.: Managing context data for smart spaces. Personal Communications, IEEE [see also IEEE Wireless Communications] 7(5), 44–46 (2000)
5. Chen, G., Kotz, D.: A survey of context-aware mobile computing research. Technical report, Dartmouth College, Hanover, NH, USA (2000)
6. de Freitas, R., Neto, B., da Graca, M., Pimentel, C.: Toward a domain-independent semantic model for context-aware computing. In: LA-WEB 2005: Proceedings of the Third Latin American Web Congress, Washington, DC, USA. IEEE Computer Society, Los Alamitos (2005)
7. Dey, A., Abowd, G., Salber, D.: A conceptual framework and a toolkit for supporting the rapid prototyping of context-aware applications. Human-Computer Interaction 16, 97–166 (2001)
8. Dey, A., Mankoff, J., Abowd, G., Carter, S.: Distributed mediation of ambiguous context in aware environments. In: UIST 2002: Proceedings of the 15th annual ACM symposium on User interface software and technology, pp. 121–130. ACM, New York (2002)
9. Dorn, C., Schall, D., Dustdar, S.: Granular context in collaborative mobile environments. In: Meersman, R., Tari, Z., Herrero, P. (eds.) OTM 2006 Workshops. LNCS, vol. 4278, pp. 1904–1913. Springer, Heidelberg (2006)
10. Intel Mobile Platform Software Development Kit 1.3 – Open Source Project for Windows and Linux (August 11, 2008), http://ossmpsdk.intel.com/
11. Formayer, H., Frischauf, C.: Extremereignisse und klimawandel in Österreich aus sicht der forschung. Technical report, WWF Austria and Institute of Meteorology, BOKU, Vienna (2004)
12. Gray, P.D., Salber, D.: Modelling and using sensed context information in the design of interactive applications. In: Nigay, L., Little, M.R. (eds.) EHCI 2001. LNCS, vol. 2254, pp. 317–336. Springer, Heidelberg (2001)
13. Henricksen, K., Indulska, J.: Modelling and using imperfect context information. In: PERCOMW 2004: Proceedings of the Second IEEE Annual Conference on Pervasive Computing and Communications Workshops, pp. 33–37 (March 2004)
14. Henricksen, K., Indulska, J., Rakotonirainy, A.: Modeling context information in pervasive computing systems. In: Mattern, F., Naghshineh, M. (eds.) PERVASIVE 2002. LNCS, vol. 2414, pp. 79–117. Springer, Heidelberg (2002)
15. Honle, N., Kappeler, U.-P., Nicklas, D., Schwarz, T., Grossmann, M.: Benefits of integrating meta data into a context model. In: PERCOMW 2005: Proceedings of the Third IEEE International Conference on Pervasive Computing and Communications Workshops, Washington, DC, USA, pp. 25–29. IEEE Computer Society, Los Alamitos (2005)
16. Kim, Y., Lee, K.: A quality measurement method of context information in ubiquitous environments. In: ICHIT 2006, pp. 576–581. IEEE Computer Society, Los Alamitos (2006)
17. Krause, M., Hochstatter, I.: Challenges in modeling and using quality of context (qoc). In: Magedanz, T., Karmouch, A., Pierre, S., Venieris, I.S. (eds.) MATA 2005. LNCS, vol. 3744, pp. 324–333. Springer, Heidelberg (2005)

18. KSOAP2 (Access Date: August 11, 2008),
    http://sourceforge.net/projects/ksoap2/
19. MXQuery (Access Date: August 11, 2008),
    http://www.dbis.ethz.ch/research/current_projects/MXQuery
20. Ranganathan, A., Al-Muhtadi, J., Campbell, R.H.: Reasoning about uncertain contexts in pervasive computing environments. Pervasive Computing 3(2), 62–70 (2004)
21. Razzaque, M.A., Dobson, S., Nixon, P.: Categorization and modeling of quality in context information. In: Proceedings of the IJCAI 2005 Workshop on AI and Autonomic Communications, Edinburgh, Scotland (August 2005)
22. Schmidt, A.: A layered model for user context management with controlled aging and imperfection handling. In: Roth-Berghofer, T.R., Schulz, S., Leake, D.B. (eds.) MRC 2005. LNCS (LNAI), vol. 3946, pp. 86–100. Springer, Heidelberg (2006)
23. Sheikh, K., Wegdam, M., van Suinderen, M.: Quality-of-context and its use for protecting privacy in context aware systems. Journal of Software 3, 83–93 (2008)
24. Tang, S., Yang, J., Wu, Z.: A context quality model for ubiquitous applications. In: IFIP International Conference on Network and Parallel Computing Workshops (NPC), pp. 282–287 (September 2007)
25. The EU WORKPAD Project, http://www.workpad-project.eu
26. Truong, H.-L., Juszczyk, L., Bashir, S., Manzoor, A., Dustdar, S.: Vimoware - a toolkit for mobile web services and collaborative computing. In: 34th EUROMICRO Conference on Software Engineering and Advanced Applications (2008)
27. Truong, H.L., Juszczyk, L., Manzoor, A., Dustdar, S.: Escape - an adaptive framework for managing and providing context information in emergency situations. In: Kortuem, G., Finney, J., Lea, R., Sundramoorthy, V. (eds.) EuroSSC 2007. LNCS, vol. 4793, pp. 207–222. Springer, Heidelberg (2007)
28. van Bunningen, A.H., Feng, L., Apers, P.M.G.: Context for ubiquitous data management. In: UDM 2005: Proceedings of the International Workshop on Ubiquitous Data Management, Washington, DC, USA, pp. 17–24. IEEE Computer Society Press, Los Alamitos (2005)
29. Weiser, M.: The computer for the twenty first century. Scientific American 265(3), 94–104 (1991)
30. Xu, C., Cheung, S.C.: Inconsistency detection and resolution for context-aware middleware support. In: ESEC/FSE-13: Proceedings of the 10th European software engineering conference held jointly with 13th ACM SIGSOFT international symposium on Foundations of software engineering, pp. 336–345. ACM, New York (2005)

A Game-Theoretic Approach to Co-operative Context-Aware Driving with Partially Random Behavior

Stefan Rass[1], Simone Fuchs[2], and Kyandoghere Kyamakya[2]

[1] Institute of Applied Informatics, System Security Group,
Klagenfurt University,
stefan.rass@uni-klu.ac.at
[2] Institute of Smart System Technologies, Transportation Informatics Group,
Klagenfurt University,
{simone.fuchs,kyandoghere.kyamakya}@uni-klu.ac.at

Abstract. We present a game-theoretic approach to co-operative driving under the assumption that drivers are willing, yet unable to follow recommendations of their adaptive driving assistance systems (ADAS) precisely. Unknown road conditions, certain health conditions, or other external impacts may render drivers unable to exactly realize their optimal behavior. We argue the importance of context information for co-operative driving, and sketch a framework where driving assistance systems communicate with each other to negotiate optimal recommendations for their specific drivers. As the final decision is nevertheless up to the driver, it is interesting to investigate what happens if drivers are unwilling or unable to follow the recommended choices. Using game theoretic tools, we derive bounds on the impact that partially random behavior can have on the own utility (safety) of a driver.

1 Introduction

Driving assistance systems (DAS) offer support in potentially dangerous driving situations. Co-operative systems improve their performance by sharing context-information with each other. Knowledge about the driving context has substantial influence on the decisions and recommendations of advanced driver assistance systems. The more a system knows about its environment, the more intelligent are the decisions it can make and pass on to the driver. But what are the negative consequences if a driver does not yield to the recommendation? Stated differently, what consequences can we expect from unskilled drivers that are willing to follow a recommendation, but cannot? Thirdly, drivers may suffer from bad health conditions, which can also inhibit precise maneuvers, but neither the skill-level or the precise condition of the driver will be available in full detail. Faulty components at the vehicle may have remained undiscovered yet, and the skill level or health condition of a driver is an information which's conveyance would surely violate a driver's privacy. Still, that information is of importance to other drivers, because actions have to be chosen differently, depending on how much

D. Roggen et al. (Eds.): EuroSSC 2008, LNCS 5279, pp. 154–167, 2008.

trust we can have in the skills of our neighbor on the road. The goal of this work is a theoretical estimation how much is lost by lack of certain information and resulting random behavior due to missing skills or bad conditions of the road, car, or the own health. Our solution implicitly accounts for privacy issues of a driver (information that is private and thus not communicated may be considered as a source of randomness by others), as well as for imprecise and uncertain information

We present an approach that estimates the consequences of diverging driving behavior exploiting methods from game theory. As this work focuses on safety, we shall assume the driving assistance system to calculate its recommendations w.r.t. maximum safety. Other goals are imaginable (such as fast travel), but are not covered here. Potential misbehavior is assumed to be random, and the goal of our model is the analysis of possible danger arising from such phenomena rather than explaining misbehavior. We emphasize that this work presents a purely theoretical concept, which's implementation is subject of currently ongoing work. We therefore will restrict ourselves to the theoretical results, and defer practical considerations to upcoming papers.

2   Related Work

Over the last decade, co-operative driving has become a very active field of research [17,20,21], indicating it as a promising and important concept for the future. However, we are not aware of any proposal of estimating the damage due to misbehavior, either intended or not. To the best of our knowledge, our proposal is the first attempt to explicitly assess the loss arising from such misbehavior, and hence can be a valuable asset in any project dealing with cooperation for increased security. A recent attempt [14] to model driving behavior under extreme conditions is not game theoretic, and as such essentially different from ours. The idea of using game theory for describing driver's behavior is not new, though: In [1], an extensive form game is used to model the driving process involving various decisions throughout a trip. The optimality of single decisions is formalized and investigated, but cooperation among several drivers is not explicitly considered.

The possible failure of Nash equilibria in accurately describing situations where humans are involved has been pointed out in [6] almost a decade ago, but unlike the approach presented here, the authors of [6] focus on certain signals that should be incorporated when looking for optimal decisions. The absence of such signals may indeed become a source of randomness in the behavior drivers may observe among each other, and our work adds to previous considerations the aspect of what happens if certain inputs cannot be made known to the driver.

In [4,27], the concept of dynamic games and dynamic coalitions has been used to develop mechanisms for driving aid and assistance systems. The approach is without doubt very interesting, but also lacks an account for possible distortions. The dynamic nature of the model may allow for application of classical stability theory in order to analyze the effects of perturbations. In contrast to this, our work provides similar results for (discrete) repeated games without needing to rely on difficult numerics involved with solutions of differential equations or maximization of complex functionals.

The use of game theory to support and analyze the driving process is further backed up by [5], whose author identifies adaptive driving assistance systems (ADAS) as a valuable asset especially for older drivers. Actions chosen by kids or old persons may occasionally appear irrational or even random, because the person may either be unaware of a risk (kids), or too slow to react appropriately on certain incidents. Each is a source of possible danger and deserves closer investigation, which we provide a starting point for in this work.

The author of [15] thoroughly discusses the problem of dynamical models for decision making, and uses the driving context (route choice models) for comparing the obtained results to recent results of experimental games in that context. The work focuses on how decisions are made and what influences are there that cause possibly bad decisions.

Other approaches include [12], in which the problem of cooperation is cast into a problem of agent synchronization, which then allows for invocation of tools from stability theory. Their approach is not game theoretic, and synchronization makes it difficult to focus on a single user's needs, since synchronization means reaching a global consensus, which may not be optimal for a single user any more, but may indeed be optimal for the whole set of users. We shall focus on single individuals here.

The authors of [2] elaborate on the importance of analyzing effects of misbehavior for cooperations using a simulation-based approach. (Mis)behavior in games is discussed in [3], presenting an approach that rests on similar assumptions as ours, but does not provide measures on the loss that individuals suffer from occasional non-rational play.

3   A Context Model

Let us sketch a context model for co-operative driver assistance systems. Within a driving situation we consider four major context areas: 1) the environment, 2) the driver, 3) the own-vehicle, and 4) the driving regulations (traffic law). The environment is by far the most complex context with the greatest variety of participating objects. It consists of the spatial context, which is basically the type of the road (highway, city, etc.) and the local context, which is a regional physical environment with specialized driving rules (intersection, level crossing, etc.). The spatial context in general is the physical environment, the own vehicle is currently driving in, so basically it gives the type of the current road. A spatial context also contains other traffic objects, which are relevant for the driving task. Examples are traffic signs, markings and traffic lights, as well as other participants. Typically the spatial context is valid for a longer timespan (from minutes up to hours). Every spatial context can be further specialized by a local context. The local context is a regional physical environment that is located within a spatial context and typically valid for a very short timespan (seconds or minutes). Special driving rules are usually applied to a local context.

Examples for a local context are intersections, zebra and level crossings, tunnels, roundabouts etc. A local context always depends on the currently valid spatial context.

Within the spatial/local context, traffic objects like signs or markings are valid and other participants (vehicles, pedestrians) are located within it. The relationships of spatial context, traffic objects and participants to the own vehicle are important to a driver assistance system. Depending on them, various driving regulations are valid and specific behaviors must be applied. Road conditions complement the driving environment.

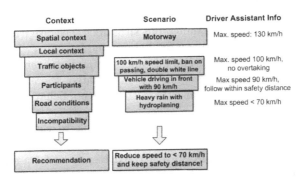

Fig. 1. Influence of context on the decision process of a driver assistance system [?]

About the driver, we would like to know the experience, the current state (fatigue, health condition, etc.) and the risk willingness. With regard to the own-vehicle, the standard parameters, like speed, and the technical equipment are important. Recommended behavior differs e.g. with the presence of an antilock brake system, because of a decreased stopping distance. Finally, the driving regulations provide the allowed and recommended behavior in a specific situation.

Figure 1 gives an example of how context influences the recommendations of a driver assistance system. The first step in the hierarchy determines the spatial context, which dictates the most general driving rules for the current environment. For example, without further restrictions the standard speed limit of 130 km/h applies for the highway. If there is no local context, we can directly switch to the presence of traffic objects. Let's assume that a 100 km/h speed limit, a ban-on-passing and a double white line are present. If the driver assistance system is aware of this restrictions it will adjust its speed and overtake recommendations. Without further participants this information will just determine the allowed legal speed and can be used, e.g. by an adaptive cruise control to adjust the current speed. When looking further, we will eventually detect another vehicle driving in front of us with 90 km/h. A context-unaware assistant without knowledge of the speed limit and the ban-on-passing would recommend to take over, whereas the context-aware system recognizes that it it now allowed to overtake and tells the driver to stay behind and adjust speed. In case of additional bad weather conditions like heavy rain, there is danger of hydro planning and the driver must be told to further slow down and increase the safety distance. This simple example already shows the high influence context-awareness has on the decisions on a driver assistance system. Reasoning starts with determining the standard behavior and all additional detected objects and conditions impose further restrictions. We refer the reader to [9] for a detailed discussion of the context model.

In a perfect world, the driver assistance system has every relevant information about the current driving environment and all drivers and participants behave perfectly legal and reasonable. Unfortunately, in the real world this is rather often not the case. Context-information is often incomplete, uncertain and ambiguous; traffic participants behave selfishly, irrationally and unforeseeable, regardless of the consequences. This leads to dangerous situations, because irrational actions of other participants (e.g. ignoring a red light or sudden stops without a reason) can not be predicted under any circumstances. However, even normal maneuvers like small changes in speed cannot be anticipated, therefore most driving assistance systems generally assume that other drivers behave continuously for the sake of simplicity.

Even the driver of the own vehicle will not be able to yield to a recommendation with 100% accuracy. For example, road conditions (e.g. black ice) that are unknown to the system could prevent the driver to perform a rapid braking maneuver. Also, if the driver is told to speed up, let's say 7 km/h, it will be difficult to perform this requested maneuver exactly - most speedometers don't even have a scale this accurate. Our approach is mainly concerned with the price that has to be paid for small or large deviations in recommended behavior, either intentionally or unconsciously performed by drivers.

Speaking about co-operative driving, we henceforth assume that driving assistance systems continuously communicate, negotiating optimal recommendations to their respective drivers in order to maximize security. Consequently, we assume the existence of a utility function determining the recommendations presented to a driver, and so the drivers can be modeled as *players* in a game where strategies accord with the driving parameters/actions (speed, distances, lane change, etc.) and utility functions are measures of safety resulting from several action configurations (see section 4.2). Because we can assume the driving parameters to be continuous quantities, strategy spaces for drivers are continuous subsets of the Euclidian vector space.

4 Preliminaries

Before presenting our results, we shortly revise the necessary foundations of game theory in this section for convenience of the reader.

4.1 Elements of Game-Theory

Consider a finite set of players (drivers), which we denote as $N = \{1, 2, \ldots, n\}$, with n possibly being very large. Notice that throughout this work, we use the words "driver" and "player" synonymously. A non-cooperative n-Person game in *normal form* is a triple $\Gamma = (N, S, H)$ composed of a set of *players* $N = \{1, 2, \ldots, n\}$, a set of compact and bounded d-dimensional *strategy spaces* $S = \{S_1, S_2, \ldots, S_n\}$, where $S_i \subset \mathbf{R}^d$ for all i, and a set of *payoff functions* $H = \{u_1, \ldots, u_n\}$ dependent on the strategies of each player, i.e. $u_i : \times_{i=1}^n S_i \to \mathbf{R}$ for all $i \in N$. Throughout this work, all games are understood to be non-cooperative, hence we may occasionally omit the explicit term "non-cooperative" and simply speak about a "game". A pure strategy is any member of the strategy set of player, as opposed to a mixed strategy, which is a probability distribution over the strategy set. We shall concentrate on mixed strategies, and their implied expected average, which we (for the sake of simplicity) again denote as $u_i(s_i, s_{-i})$, with the standard notational convention that s_i is the mixed strategy chosen by the i-th player, and s_{-i} is the set of mixed strategies chosen by i's opponents.

If players act selfishly while paying attention to their opponents actions, any resulting stable situation will be a Nash-equilibrium [23,16]: The vector of strategies $s^* = (s_1^*, \ldots, s_n^*)$ is called a *Nash-equilibrium* for the non-cooperative game $\Gamma = (N, S, H)$, if every player i chooses a strategy s_i^* which gives maximum utility, i.e. if

$$u_i(s_i^*, s_{-i}^*) \geq u_i(s_i, s_{-i}^*) \quad \forall i, \forall s_i \in S_i. \tag{1}$$

It is well known that such equilibria exist if the strategy spaces are finite, however, we will need a more general result that has been given a few years after Nash's existence theorem:

Theorem 1 ([11,13]). *If for a game in normal form, the strategy spaces S_i are nonempty compact subsets of a metric space, and the payoff functions are continuous, then there exists at least one Nash-equilibrium in mixed strategies.*

Many results in game theory presume fully rational behavior of players. Our setup assumes a mechanism of advice that tells a driver an optimal choice, which cannot be implemented with full precision. Consequently, although there may be no intention behind it, drivers may still act different to a rational expectation, due to technical limitations of the vehicle, or unknown conditions of the road preventing exact driving. In a more general context, we may summarize this as partial randomness or irrationality, since we may have a quite good idea what to do, but cannot exactly realize the strategy. Hence, we refer to this as *acting semi-rational*. The now introduced concept of games with semi-rational players serves as mathematical model for such behavior:

A non-cooperative n-Person game with *semi-rational players* arises from a classical non-cooperative game by endowing the set of players with probability distributions reflecting the uncertainty of strategy choices. This is made precise in the following definition:

Definition 1 (Game with semi-rational players). *Let $\Gamma = (N, S, H)$ be a game in normal form. A non-cooperative game with semi-rational players is a quadruple $\Gamma = (N, F, S, H)$, where $F = \{F_1, \ldots, F_n\}$ is a set of probability measures of random variables X_i over a metric space $\mathcal{S}$. Each random variable*

X_i satisfies $\mathrm{E}(X_i) = 0$; the set $\mathcal{S}$ is assumed to be a continuum and $S_i \subseteq \mathcal{S}$ for all strategy spaces S_i.

The independent random variables $X_1, \ldots, X_n$ reflect the errors each player inevitably makes when trying to realize a chosen strategy, i.e. if s_i^* is the optimal strategy for player i, then the strategy actually played is $s_i^* + X_i$, where $X_i \sim F_i$. The property $\mathrm{E}(X_i) = 0$ models the assumption that players still perform optimal in the long run average, but although a fully rational strategy is attempted, no player can advance the own skills beyond some personal and/or technical limit. Players who disregard any contextual information can be considered as playing with different utility functions, so this causes no conflict with our definition. Naturally, this relates our setting to *perturbed* and *disturbed* games [25], but contrary to perturbed games (and their resulting equilibria, referred to as *perfect equilibria*), solutions of games with semi-rational players are not characterized through a limit process of completely mixed strategies (cf. [25] and Lemma 1). In a disturbed game, an opponent cannot observe the utility, but a perturbed version of it which is $u + X$ for some utility u and some random variable X. Contrary to this, we assume the strategies to be imprecise. Although this could technically be viewed as a disturbed game, our approach yields substantially simpler proofs of according results.

It remains to be clarified what we mean by an *equilibrium* of a game with semi-rational players: Since we cannot hope to directly apply Nash's definition of an equilibrium, we need to cope with the randomness first. Thanks to the assumption of zero mean for each random variable X_i, this is easy, as Nash's definition then transforms as follows: According to the defining property of a Nash-equilibrium (1), an equilibrium situation is present if $u_i(s_i^*, s_{-i}^*) \geq u_i(s_i, s_{-i}^*) \quad \forall i, \forall s_i \in S_i$. Replacing the optimal strategy s^* by an imprecise strategy $s_i^* + X_i$, the terms become random, and we can ask for their long-run average, which is $\mathrm{E}_{X_i}(u_i(s_i^* + X_i, s_{-i}^* + X_{-i}))$. Accordingly, we define an equilibrium for a game with semi-rational players by being optimal in terms of expected utility:

Definition 2 (equilibrium for games with semi-rational players). *Let $\Gamma = (N, F, S, H)$ be a non-cooperative game in normal form with semi-rational players, then a strategy s^* is called an* equilibrium, *if the expected utility of each player is optimal under the chosen strategy s_i^*, i.e. if $\forall i, \forall s_i \in S_i$,*

$$\mathrm{E}_{X_i}(u_i(s_i^* + X_i, s_{-i}^* + X_{-i})) \geq \mathrm{E}_{X_i}(u_i(s_i + X_i, s_{-i}^* + X_{-i})). \qquad (2)$$

There is a natural relation to the concept *correlated equilibria* [16], where players derive their strategies from correlated random variables. This, however, is not the case in our setting, since we assume randomness of strategies to be uncorrelated among players, but their choices to be correlated through the utility function. The latter is the case in any game, but unlike correlated equilibria, imprecisions of actions of a driver are independent of the deviations other drivers exhibit. Regarding the existence of equilibria, we have the following result:

Lemma 1. *For every game with semi-rational players, there exists a normal form game with the same set of equilibria.*

Proof. Define the random variable $Y_i := -X_i$, then Y_i has the density $g_i(x) := f_i(-x)$, by a change of variables [18]. With $s = (s_i, s_{-i})$, re-write the right-hand side of (2) as $\mathrm{E}(u_i(s_i^* + X_i, s_{-i}^* + X_{-i})) = \mathrm{E}(u_i(s_i^* - Y_i, s_{-i}^* - Y_{-i})) = \mathrm{E}u_i(s - Y_i)$. The latter term is equal to $\mathrm{E}u_i(s - Y_i) = \int_{S_i} u_i(s - y)f_i(-y)dy = u_i * g_i$. By rewriting the equilibrium condition (2) in terms of $\tilde{u}_i := u_i * g_i$, we obtain $\tilde{u}_i(s_i^*, s_{-i}^*) \geq \tilde{u}_i(s_i, s_{-i}^*)$ $\forall i, \forall s_i \in S_i$, which is the defining condition of Nash-equilibrium (cf. Equation (1)). It follows that a situation s^* is an equilibrium for a given game $\Gamma = (N, F, S, H)$ with semi-rational players if and only if it is a Nash-equilibrium of the game $\Gamma' = (N, S, H')$ in normal form with $H' := \{u_i * g_i | g_i(x) = f_i(-x), f_i \in F\}$.

Using Theorem 1, this lemma instantly establishes the following existence result, which ensures that solutions to games with semi-rational players exist if standard conditions are satisfied. The hypothesis under which solutions exist is similar to the one of Theorem 1 with an additional constraint on the measures:

Theorem 2. *Every game with semi-rational players with nonempty compact strategy spaces, continuous payoff functions and compactly supported uncertainty measures has at least one equilibrium.*

Knowing that there are equilibria situations, i.e. solutions to the game we have created, we can ask for how much those equilibria deviate from a situation that would arise in a game with perfect players. This is subject of Section 5.

4.2   A Measure of Safety

Our presentation in this work will follow the ideas of [10] to define an appropriate utility function measuring safety of a driving process: Assume that from historical data, the ADAS records deviations between its recommendations and the true behavior of a driver. With that information, we can set up an empirical distribution which allows for simulations of scenarios under different possible behaviorial alternatives of a driver. Considering overtaking as an example, we may run a number of simulations each of which simulates the driver's behavior according to the randomness we have recorded from past experience. The relative frequency of successful completions of the maneuver (where any situation that does not lead to any physical contact of the vehicle with another object may be considered as a success) is then the utility we can expect from our recommendation. The utility then has the interpretation of the probability of successful completion of a maneuver, and can even be presented to a driver.

As our main concern is safety of a driver, let us focus our attention on only two possible outcomes of a decision, which are

1. no accident.
2. accident with possible injuries or damage to the vehicle.

Note that here, we do not explicitly treat differences between light injuries and severe injuries up to paralysis. Neither do we distinguish between damages like scratches in the lacquer or a broken engine. We treat them all on an equal basis as giving zero utility, while avoiding an accident is assigned utility 1. The reverse assignment of utility (corresponding to loss) is known as 0-1-loss in decision theory (cf. [24]). Let us define our utility as binary valued variable depending on a choice s:

$$u(s) = \begin{cases} 0, & \text{if } s \text{ leads to an accident;} \\ 1, & \text{otherwise.} \end{cases}$$

Let us illustrate our thoughts with an example: Consider an overtaking maneuver on two lanes on a highway, involving three vehicles as shown in figure 2.

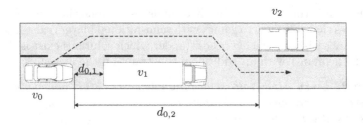

Fig. 2. Overtaking maneuver involving three vehicles

Vehicles are termed 0,1,2 (from left to right) and drive at the corresponding velocities v_0, v_1 and v_2. The distance $d_{0,2}$ is assumed 90m and the distance $d_{0,1}$ is 30m in our example. Initially, assume the velocities to be exactly known as $v_0 = 100km/h, v_1 = 100km/h$ and $v_2 = 120km/h$ (assuming a standard situation on a German highway). Then a simple calculation shows that the overtaking maneuver can be done by accelerating to 130km/h without risk (i.e. utility comes to 1). Keeping the strategy s to be "accelerate to 130", on the other hand, if vehicle 2 goes only with 100km/h, then a crash will surely occur (utility is 0). Now, suppose that either from observation or from communication (cooperation), we know that the driver of vehicle 2 maintains his speed with a standard deviation of 10 (assuming a normal distribution of speed choice around the optimal (expected) value of 120km/h), which makes 99.7% of the random choices lie within the interval $90 \leq v_2 \leq 150$. There is a small chance that the driver of vehicle 2 slows down and thus creates a dangerous situation if the overtaking maneuver is carried out with the recommended choice. Drawing random samples and recalculating the scenario for a large number of cases then gives a relative frequency of times when the maneuver cannot be completed successfully. Running a simulation with 10000 trials gives 930 cases in which a crash occurs, which comes to a probability of 0.093 for a crash. The utility value $u = 1 - 0.093 = 0.907$ is then presented to the driver as likelihood of successful completion of the maneuver if he does the overtaking.

5   Bounding the Loss

We would like to obtain some measure giving the deviation of Nash-equilibria when moving from exact play to imprecise strategies. Let a game with semi-rational players be given, and call $\Gamma(F) = (N, S, H)$ the corresponding game according to Lemma 1.

Our main goal in this section is establishing a measure of how much is lost by occasionally irrational play. Since the played strategies are inherently random, we shall provide bounds on the probability to loose more than some given threshold by the imprecision of the player's actions. This can be answered on probability theoretic grounds as follows, using Theorem 3, better known as *McDiarmid's inequality*, as a technical tool, which will later establish Theorem 4:

Theorem 3 ([22, Lemma 1.2]). *Let $X_1, \ldots, X_n$ be independent random variables, where X_k takes values in a set A_k. Suppose that the measurable function $f : \prod A_k \to \mathbf{R}$ satisfies $|f(x) - f(x')| \leq c_k$, for some constants c_k whenever x, x' differ only in the k-th coordinate. Let Y be the random variable $f(X_1, \ldots, X_k)$, then for any $\varepsilon > 0$, $\Pr\{|Y - \mathrm{E}(Y)| \geq \varepsilon\} \leq 2\exp\left(-2\varepsilon^2 / \sum_{i=1}^{k} c_k^2\right)$.*

Theorem 4. *Let Γ be a game with semi-rational players. If the payoff functions are continuous and convex in each strategy, and the strategy spaces are compact, then for any $\varepsilon > 0$, the probability of deviating more than ε from the optimal utility is exponentially small. More precisely,*

$$\Pr\left\{\left|u_i(s_i^*, s_{-i}^*) - u_i(s_i^* + X_i, s_{-i}^*)\right| < \varepsilon\right\} \geq 1 - 2\exp\left(\frac{-\varepsilon^2}{n\,\|u_i\|_\infty^2}\right), \qquad (3)$$

for $s_i + X_i$ being the randomly perturbed strategy of the i-th player.

Proof. Let $s^* = (s_1^*, \ldots, s_n^*)$ be an equilibrium situation according to definition 2. Call $Y_i := s_i^* + X_i$ the random variable that reflects player i's behavior. By the convexity of u and Jensen's inequality [7],

$$\mathrm{E}u_i(Y_i, s_{-i}^*) \geq u_i(\mathrm{E}Y_i, s_{-i}^*) = u_i(s_i^*, s_{-i}^*) \geq u_i(Y_i, s_{-i}^*), \qquad (4)$$

where the second inequality follows from the definition of an equilibrium (1). Notice that within the chain of inequalities, the optimal utility $u_i(s_i^*, s_{-i}^*)$ lies somewhere between $\mathrm{E}u_i(Y_i, s_{-i}^*)$ and $u_i(Y_i, s_{-i}^*)$. Abbreviating the (real-valued) terms in expression (4) as $a := \mathrm{E}u_i(Y_i, s_{-i}^*), b := u_i(\mathrm{E}Y_i, s_{-i}^*), c := u_i(Y_i, s_{-i}^*)$, (4) can be re-written as $a \geq b \geq c$, from which we conclude $|b - c| \leq |a - c|$. For $\varepsilon > 0$, let us introduce the events $A := \{|a - c| < \varepsilon\}$ and $B := \{|b - c| < \varepsilon\}$, then the last inequality implies that $A \subseteq B$, and by taking probabilities $\Pr(A) \leq \Pr(B)$. Calculating the probabilities for the complementary events (i.e. multiplying the inequality with -1 and adding 1) gives

$$\Pr\{|a - c| \geq \varepsilon\} \geq \Pr\{|b - c| \geq \varepsilon\}, \qquad (5)$$

in which we can upper-bound the right term using McDiarmid's inequality. Because u_i is continuous on a compact set, we have $\|u_i\|_\infty = \sup_s |u_i(s)| < \infty$, so by Theorem 3, $2\exp\left[-\varepsilon^2/(n\|u_i\|_\infty^2)\right] \geq \Pr\{|a - c| \geq \varepsilon\}$, where n is the dimension of the full strategy vector (s_i^*, s_{-i}^*), and therefore by (5), $\Pr\{|b - c| \geq \varepsilon\} \leq 2\exp\left[-\varepsilon^2/(n\|u_i\|_\infty^2)\right]$, which establishes the claim by taking complementary probabilities again.

It is often observable that an increase of speed is not necessarily tied to a proportional saving of time. In some sense, if we take travel time as utility and speed as a strategy, then we can only increase the speed up to its legal limit, while still needing to obey the other drivers behavior. If everyone tries to maximize the speed, then we are back at the situation where drivers follow similar speed patterns, in which case nobody can significantly win over another. This intuitively indicates that the travel time has a minimum somewhere in the middle, and the assumption of convex utility intuitively makes sense for this example. Yet, we shall relax the assumption at the cost of loosing the global nature of the theorem in the following.

Discussing the case of non-convex payoffs is worthwhile, as the convexity assumption may be violated by many real-world applications. Unfortunately, the proof of Theorem 4 vitally relies on that assumption, but we can show that the result still holds "locally" by virtue of an approximation argument. We base our argument on the following well-known result, belonging to the class of *approximation theorems*:

Theorem 5 ([19, pg. 320]). *Let $(\delta_k)_{k\in\mathbf{N}}$ be a Dirac-sequence. Then the sequence $(f * \delta_k)$ is L^1-convergent towards f for every function $f \in \mathcal{L}^1(\mathbf{R}^n)$. Furthermore, if f is uniformly continuous and bounded on $\mathbf{R}^n$, then the convergence $(f * \delta_k) \to f$ is uniform.*

Lemma 2. *Let u be a continuous function on a compact set S, then for every $\varepsilon > 0$ and every point $s \in S$, we can find a convex function $\widehat{u}$ such that $|\widehat{u}(s') - u(s')| < \varepsilon$ for all s' in some neighborhood of s.*

Proof. In the light of Theorem 5, we can use of smoothing techniques based on Dirac-sequences as follows: Let $(\delta_k)_{k\in\mathbf{N}} \in C^1(S)$ be a Dirac-sequence. Continuity of u on a compact set implies uniform continuity and $u \in L^1(S)$, hence the second statement Theorem 5 applies: For any $\varepsilon > 0$, we can find an $n > 0$ such that $\|u - u * \delta_n\|_\infty < \frac{\varepsilon}{2}$. By [26, Lemma 7.22], $u * \delta_n$ is everywhere differentiable. Fix a point $s \in S$ and call $\widehat{u}(x) := [(D(u * \delta_k))(s)](x - s) + (u * \delta_k)(s)$ the tangent hyperplane at $(s, (u * \delta_k)(s))$, which is obviously convex. Moreover, we can find $\eta > 0$ such that $\|u * \delta_n - g\|_\infty < \frac{\varepsilon}{2}$ on $B_\eta(s)$, which finally gives $|u(s') - \widehat{u}(s')| < \frac{\varepsilon}{2} + \frac{\varepsilon}{2}$ on the open ball $B_\eta(s)$.

The argument of Lemma 2 is graphically displayed in Figure 3. The dashed lines mark the ε-neighborhood of some utility function u, which is convex and not differentiable. The approximation via the tangent of some smoothed version of u is carried out in a neighborhood of the origin.

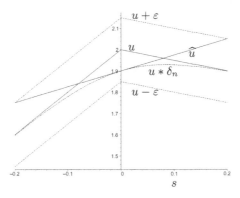

Fig. 3. Local approximation of a general function u via a convex function $\widehat{u}$ within a uniform ε-bound

Thanks to this lemma, we can trade the convexity assumption for a local generalization of Theorem 4, which we state as

Theorem 6. *Let Γ be a game with semi-rational players. If the payoff functions are continuous, and the strategy spaces are compact, then for any $\varepsilon > 0$ and any equilibrium $s^* \in S$, there exists some number $\eta(s) > 0$, such that the probability of locally deviating more than ε ($0 < \varepsilon < \eta(s)$) from the optimal utility is exponentially small. More precisely,*

$$\Pr\left\{\left|u_i(s_i^*, s_{-i}^*) - u_i(s_i^* + X_i, s_{-i}^*)\right| < \varepsilon\right\} \geq 1 - 2\exp\left(\frac{-2\varepsilon^2}{nC^2}\right), \qquad (6)$$

for $s_i + X_i$ being the randomly perturbed strategy of the i-th player, and C being a constant that only depends on the function u restricted to the neighborhood $B_{\eta(s)}(s)$.

Proof. For a chosen point s, call $\widehat{u}$ the convex uniform $\frac{\varepsilon}{2}$-approximation of u in a neighborhood $B_\eta(s)$ for some $\eta > 0$, according to Lemma 2. Within $B_\eta(s)$, the hypothesis of Theorem 4 is satisfied with $\frac{\varepsilon}{2}$. The role of $\|u_i\|_\infty$ in (3) is played by the constant C, found as $C = \sup_{s_1, s_2 \in B_{\eta(s)}(s)} |\widehat{u}(s_1) - \widehat{u}(s_2)|$, and the assertion follows (cf. Theorem 3). $\square$

The statements of Theorem 6 and Theorem 4 can be interpreted as follows: The probability of large deviations is exponentially small, meaning that sudden abrupt changes in utility (e.g. crashes) are unlikely to occur if drivers are willing to follow recommendations. This in some sense rules out sudden "chaotic" behavior exhibited by interacting drivers, but only up to a certain probability. Here, the term chaotic is also not to be understood in the classical sense of dynamic systems theory, rather we mean unexpected events that are significantly more dangerous than the situation was before that event.

6   Conclusion

We have presented a game-theoretic framework that allows for a simple analysis of possible danger under the assumption that drivers interact non-rational on certain occasions. We assume co-operative adaptive driving assistance available for each driver, which's decisions undergo random perturbations due to conditions beyond the drivers control. These occasionally random behavior leads to a decrease of safety, or utility in general, and our work focuses on how to quantify this loss. Our results allow for the conclusion that small uncertainties in the behavior of drivers may unlikely yield severe safety problems. The focus of the present analysis lies on the quantification of loss that is experienced *subjectively*, i.e. by a single user and the own actions. Analyzing the global view and the loss of social welfare under simultaneous random behavior of participants is subject of future research, as well as evaluating the theoretical framework under real-world conditions with real drivers.

Our framework is general, and as such applicable to a perhaps much wider range of applications, not necessarily limiting ourselves to the analysis of road traffic. Future work in this direction includes improving our yet probabilistic bounds to more tight ones that may take into account further information about other drivers, like current (health) condition, as far as this information can be made available. We believe that our results can serve as a valuable add-on for theories dealing with co-operative driving as such, as our model is general and therefore applicable to a considerable range of applications, even wider than the field of co-operative driving.

References

1. Aumann, R.J., Hart, S., Perry, M.: The absent-minded driver. Games and Economic Behavior 20, 102–116 (1997)
2. Bissey, M.-E., Ortona, G.: A simulative frame to study the integration of defectors in a cooperative setting. P.O.L.I.S. department's Working Papers, Department of Public Policy and Public Choice - POLIS (January 2002)
3. Camerer, C.F., Ho, T., Chong, K.: Behavioral game theory: Thinking, learning and teaching (November 2001); Caltech Working Paper
4. Chin, H.H., Jafari, A.A.: Dynamic coalition game for driver associate. In: 12th International Symposium on Dynamic Games and Applications, July 3–6 (2006)
5. Davidse, R.J.: Assisting the older driver: intersection design and in-car devices to improve the safety of the older driver. PhD thesis, University of Groningen (2007)
6. Eckel, C., Wilson, R.K.: The human face of game theory: Trust and reciprocity in sequential games. In: Ostrom, E., Walker, J. (eds.) Trust and Reciprocity: Interdisciplinary Lessons from Experimental Research, ch. 9, pp. 245–274. Russell Sage Foundation, New York (2003)
7. Elstrodt, J.: Maß- und Integrationstheorie, 3rd edn. Springer, Heidelberg (2002)
8. Fuchs, S., Rass, S., Lamprecht, B., Kyamakya, K.: Context-awareness and collaborative driving for intelligent vehicles and smart roads. In: 1st International Workshop on ITS for an Ubiquitous ROADS, July 2007. IEEE Press, Los Alamitos (2007)

9. Fuchs, S., Rass, S., Lamprecht, B., Kyamakya, K.: A model for ontology-based scene description for context-aware driver assistance systems. In: 1st International Conference on Ambient Media and System, Quebec City, CA, February 2008, pp. 1–8. ACM, New York (2008)

10. Fuchs, S., Rass, S., Kyamakya, K.: Constraint-based context-rule representation and risk classification for driver assistance systems. In: The First Annual International Symposium on Vehicular Computing Systems (2008)

11. Fudenberg, D., Tirole, J.: Game Theory. MIT Press, London (1991)

12. Ghabcheloo, R., Aguiar, A.P., Pascoal, A., Silvestre, C.: Synchronization in multi-agent systems with switching topologies and non-homogeneous communication delays. In: 46th IEEE Conference on Decision and Control, 2007, 12-14 December 2007, pp. 2327–2332 (2007)

13. Glicksberg, I.L.: A further generalization of the Kakutani fixed point theorem, with application to nash equilibrium points. In: Proceedings of the American Mathematical Society, vol. 3, pp. 170–174 (February 1952)

14. Hamdar, S.H.: Towards modeling driver behavior under extreme conditions. Master's thesis, Faculty of the Graduate School of the University of Maryland (2004)

15. Helbing, D.: Dynamic decision behavior and optimal guidance through information services: Models and experiments. In: Schreckenberg, M., Selten, R. (eds.) Human Behaviour and Traffic Networks, pp. 47–95. Springer, Berlin (2004)

16. Holler, M.J., Illing, G.: Einführung in die Spieltheorie. Springer, Heidelberg (2003)

17. Kais, M., Bouraoui, L., Morin, S., Porterie, A., Parent, M.: A collaborative perception framework for intelligent transportation systems applications. In: World Congress on Intelligent Transport Systems (November 2005)

18. Klebaner, F.C.: Introduction to stochastic calculus with applications, 2nd edn. Imperial College Press, London (2005)

19. Königsberger, K.: Analysis 2, 5th edn. Springer, Berlin (2004)

20. Luo, J., Hubaux, J.-P.: A survey of inter-vehicle communication. Technical Report IC/2004/24, School of Computer and Communication Sciences EPFL, CH-1015 Lausanne, Switzerland (2004)

21. Marsden, G., McDonald, M., Brackstone, M.: Towards an understanding of Adaptive Cruise Control. In: Transportation Research, Part C, Emerging Technologies, vol. 9, pp. 33–51 (2001)

22. McDiarmid, C.: On the method of bounded differences. In: Surveys in Combinatorics, Cambridge [u.a.]. Lect.Note Ser. 141, pp. 148–188 (1989)

23. Nash, J.F.: Non-cooperative games. Annals of Mathematics 54, 286–295 (1951)

24. Robert, C.P.: The Bayesian choice. Springer, New York (2001)

25. van Damme, E.: Stability and perfection of Nash equilibria. Springer, London (1991)

26. Walter, W.: Analysis 2, 4th edn. Springer, Heidelberg (1995)

27. Zheng, Z., de Sousa, J.B., Girard, A.: Differential game based safe controller design for intelligent cruise control. In: 44th IEEE Conference on Decision and Control, 2005 and 2005 European Control Conference. CDC-ECC 2005, December 12-15, 2005, pp. 6716–6721 (2005)

Mobile Context-Addressable Messaging with DL-Lite Domain Model[*]

Michal Koziuk, Jaroslaw Domaszewicz, Radoslaw Olgierd Schoeneich, Marcin Jablonowski, and Piotr Boetzel

Institute of Telecommunications, Warsaw University of Technology
ul. Nowowiejska 15/19, 00-665, Warsaw, Poland
{mkoziuk,domaszew,rschoeneich}@tele.pw.edu.pl,
mjablonowski@post.pl, pboetzel@wp.pl

Abstract. This paper describes an architecture for ontology based context modeling on mobile devices. The combination of the DL-Lite logic, Manchester OWL Syntax, a customized implementation of the JENA API, and an off-the-shelf relational database offers an environment which makes it possible to use ontologies on a modern mobile device. We use such form of context modeling as the foundation of a Context-Addressable Messaging service for mobile ad-hoc networks. This paper covers the details of the service as well as the internal architecture for handling context and resolving context-based addresses. Numerical performance results measured on the Nokia N800 device are provided.

1 Introduction

In mobile context-aware systems, the notion of the context of a node (mobile device) is used to describe the environment the node is located in, its current location, the role of its user, etc. Typically, context data are associated with a particular node in the network. We propose to make special use of these data to provide support for a Context-Addressable Messaging (CAM) service in mobile ad-hoc networks[1]. This service allows sending messages in which the recipients are specified by defining criteria on their context.

An example of usage of such a service is to send messages to all Firefighters located in Station Jussieu. In this example the address of the message would be the statement 'is a Firefighter and is located at Station Jussieu'. We call such a statement a Context-Based Address, as it is used to describe the destination nodes[2] through their context. In this example, the relevant context data of destination nodes are their role and location.

The main feature of the proposed CAM service is that the criteria on the context of destination nodes are specified by the message sender. This is different

---

[*] This work was supported by the 6FP MIDAS IST project, contract no. 027055.

[1] The CAM service could also be applied in other types of networks.

[2] We identify the node (device) with its user. The same identification applies if a device is associated with a place or an object.

D. Roggen et al. (Eds.): EuroSSC 2008, LNCS 5279, pp. 168–181, 2008.

from well known publish-subscribe systems, where the recipients have to express their interest in specific types of data.

The CAM service is implemented as a middleware layer which should be preinstalled on a mobile device in order to supply this service to running applications. An important underlying assumption for this service is that all network nodes share a common Domain (Context) Model. We decided to use an ontology as the Domain Model. This ontology should model the environment of deployment of the CAM service (i.e. the domain), such as, for example, emergency situations or sport events. The main benefit of using data structured according to an ontology is that reasoning can be performed on top of such data. An external Domain Model needs to be prepared and provided together with the middleware (it can be exchanged for different deployment domains) as illustrated in Figure 1.

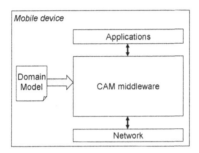

Fig. 1. Domain customization of the CAM middleware

Figure 2 presents a conceptual architecture of the CAM service as described in this paper. Given the Domain Model in the form of an ontology TBox (first row in the figure), the context of each user can be expressed in terms of the ABox of this ontology (the second row). While the TBox is the same for all nodes, the ABox is node-specific (it contains data describing the context of the node holding it[3]). Among ABox instances, we distinguish a special instance that will represent the node itself. It is the assertions related to this special instance that will be crucial for delivery of Context-Addressed Messages. We will refer to this instance as *thisNode*.

The address of a Context-Addressed Message (i.e., a Context-Based Address) is either an existing class from the ontology, or (more often) it is a definition of a new class (as shown in the fourth row of Figure 2). Through reasoning, the instances from the ontology ABox can be identified as belonging (or not) to the new class. If it happens that the *thisNode* instance belongs to the address class, the message will be delivered to the application running on top of the middleware.

Reasoning which takes advantage of the domain model ontology is one of the main advantages of CAM messaging. For example, a message sent to the address

---

[3] The amount of available context information depends on the context acquisition mechanisms used.

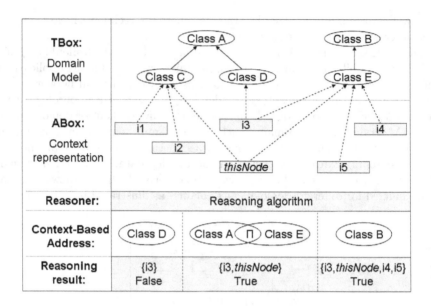

Fig. 2. CAM middleware context representation and addressing mechanism

'class B' will be delivered to every node whose ABox looks as the one shown in the example in Figure 2. Given that the *thisNode* instance belongs to class E and class E is a subclass of class B, the reasoning algorithm infers that the *thisNode* instance belongs to class B, which makes the given node an addressee for messages addressed to 'class B'.

To implement the architecture described above on a mobile device, the following elements need to be identified: (a) an ontology language with suitable expressivity for the representation of context, (b) the structure for Context-Based Addresses, (c) an efficient reasoning algorithm for address resolving, (d) a routing protocol capable of handling Context-Based Addresses. All of the above should take into account resource limitations of a mobile device.

In our first experiment with CAM [1], we tried to use off-the-shelf solutions: the OWL-DL [2] ontology language, and the Pellet [3] reasoner accessed through the JENA API [4]. These choices did not provide satisfactory performance. This paper presents a CAM service based on DL-Lite [5] ontologies, LOnt library (our own implementation of the JENA API for mobile devices), and a dedicated Context-Based Routing protocol. Note that Context-Based Routing[6] will be described in a separate publication.

This paper is organized as follows: Section 2 describes how DL-Lite is used on a mobile device for context modeling and addressing. Section 3 describes the architecture of the CAM service. Section 4 presents Nokia N800 performance results while Section 5 gives an overview of related work. Conclusions are presented in Section 6.

2   DL-Lite as Domain Modeling Language for CAM

The domain modeling language for Context-Addressable Messaging has to be balanced in terms of expressivity versus complexity. The more complexity is allowed in the model, the higher level of details can be captured. This, however, leads to an increase in the processing complexity of the model, which can very quickly make the system unusable on mobile devices.

To achieve a good performance of the CAM service, we have chosen the DL-Lite[5] logic for domain modeling. This logic is a subset of OWL-DL. DL-Lite was chosen to offer a high performance reasoning algorithm which can infer "in polynomial time with respect to [...] the size of the ontology" [7]. An additional advantage of DL-Lite is that it is mentioned as one of the few logics which can be implemented on top of a relational database and are a "subset of OWL 1.1" [7].

The DL-Lite logic can contain statements about subsumption, disjointness, role-typing, participation and non-participation constraints, and functionality restrictions (see Figure 3).

| Basic concepts: | $B := A \mid \exists R \mid \exists R^-$ |
| General concepts: | $C := B \mid \neg B \mid C_1 \sqcap C_2$ |

| Inclusion assertions: | $B \subseteq C$ |
| Functionality assertions: | $(\mathit{funct}\,R), (\mathit{funct}\,R^-)$ |

Fig. 3. DL-Lite TBox constructs [5]

CAM makes use of DL-Lite conjunctive queries [5] as Context-Based Addresses. When a node receives a Context-Addressed Message, the conjunctive query represented by the Context-Based Address of that message is evaluated. The conjunctive query is executed on top of the context storage relational database (ABox) to determine if the special instance *thisNode* is returned as a result of this query. For a detailed explanation of the conjunctive query mechanism refer to [5].

We now explain how conjunctive queries are used as Context-Based Addresses. Figure 4 presents how different atoms of a conjunctive query are mapped to message destinations. For example, using an address in the form of $B(x)$ where B denotes the class **Parent** (see the first row in Figure 4) we can address all nodes whose context defines them as 'parents' (the instance *thisNode* is returned as a result of evaluating the conjunctive query $Parent(x)$). Another type of an address could be composed of the atom $R(x, i)$ (see the fourth row in Figure 4) where R would be the relation **hasParent** and i would be the instance representing **Tom Black**. This way, using the $hasParent(x, "TomBlack")$ conjunctive query, we can address 'all the children of Tom Black'.

Addressing based on conjunctive queries can be fully appreciated when address atoms are combined together in a conjunction. In principle, it is possible to use an arbitrary number of atoms to describe the destination nodes.

Address atom	Message goes to every node that
B(x)	Belongs to class **B**. **Example:** Address: Parent(x) Destination: All persons who are a Parent.
R(x,_)	Has object property **R** assigned. **Example:** Address: hasParent(x,_) Destination: All persons that have a Parent.
R(_,x)	Is the value for the object property **R**. **Example:** Address: hasParent(_,x) Destination: All persons that are a Parent of any other person.
R(x,i)	Has object property **R** assigned, with the value equal to instance i. **Example:** Address: hasParent(x,"Tom Black") Destination: All persons whose Parent is Tom Black.
R(i,x)	Is the value for the object property **R** assigned to instance i. **Example:** Address: hasParent("Emily Black",x) Destination: All persons who are the Parent of "Emily Black".
R(x,y)	Has object property **R** assigned, with the value equal to one of the instances returned by the part of the address described by **y**. **Example:** Address: hasParent(x,y) ⊓ Manager(y) Destination: All persons who have a Parent who is a Manager.
R(y,x)	Is the value for the object property **R** assigned to one of the instances returned by the part of the address described by **y**. **Example:** Address: hasParent(y,x) ⊓ Manager(y) Destination: All persons who are a Parent of a person who is a Manager.

Fig. 4. Atoms of conjunctive queries as Context-Based Addresses

An example of a Context-Based Address constructed as a conjunction of atoms can be as follows: 'Firefighters located in Jussieu Station, in the rank of a captain'. This address imposes three conditions on the context of destination nodes: (a) they have to belong to the class **Firefighter** (atom type $B(x)$), (b) they have to have the role **isLocatedAt** with the value equal to the **Jussieu Station** instance (atom type $R(x,i)$), (c) they have to belong to the class **Captain** (atom type $B(x)$). The Context-Based Address constructed of these three atoms would look as follows:

$$Firefighter(x) \sqcap isLocatedAt(x, "JussieuStation") \sqcap Captain(x)$$

Another example could be: 'Firefighters located in a vehicle which is driving towards station Jussieu'. This address uses the $R(x,y)$ atom type as it needs to provide constraints both on the node which is target of the messages (atoms $Firefighter(x)$ and $isLocatedAt(x,y)$), as well as on the place in which that node is currently located (atoms $isDrivingTowards(y, "JussieuStation")$ and $Vehicle(y)$). The whole address would look as follows:

$$Firefighter(x) \sqcap isLocatedAt(x,y) \sqcap Vehicle(y) \sqcap$$
$$isDrivingTowards(y, "JussieuStation")$$

To implement a DL-Lite based architecture for Context-Addressable Messaging on mobile devices, the following issues, addressed in the next section, arise:

- Creating DL-Lite ontologies. Multiple convenient tools exist for development of OWL ontologies such as Protg [8]; however Protg does not provide support for DL-Lite ontologies.
- Representing and distributing a DL-Lite ontology. A standardized format that is compact and easy to parse is required.
- Using a DL-Lite ontology on mobile devices. A library capable of parsing, and handling DL-Lite ontologies on mobile devices is needed.

3   DL-Lite Based CAM Architecture

Figure 5 presents the architecture of tools required to prepare ontologies for the CAM service (design-time) as well as the architecture of the middleware installed on a mobile device required to enable the CAM service (run-time). The left hand side of the figure presents a plugin to the Protg ontology editor which is used to create DL-Lite ontologies for CAM. The right hand side of the figure shows the CAM middleware which is responsible for handling the DL-Lite ontology on a mobile device.

To facilitate creating DL-Lite ontologies we have developed a plugin to the popular Protg[8] ontology editor. This plugin is capable of reducing OWL-DL ontologies to DL-Lite ontologies, by removing those OWL-DL constructs, which are not supported by the DL-Lite logic. The information loss which occurs during the reduction process could probably be mitigated if OWL-DL specific (non-DL-Lite) constructs could be replaced by "approximating" DL-Lite constructs (rather than being simply removed). We see this as an interesting issue for future research.

Due to mobile platform resource limitations, the ontology file should be easy to parse. Parsing ontology files encoded in RDF/XML syntax is time consuming and requires a complex parser (see [9] for more details). Therefore, we have chosen the Manchester OWL Syntax [10] for representing the domain model ontology. Table 1 shows a comparison of the Manchester OWL Syntax and RDF/XML syntax. The Manchester OWL Syntax offers the following advantages: (a) it is easy to parse, (b) it is designed for human-readability, (c) DL-Lite ontology files encoded in the Manchester OWL Syntax turned out to be approximately twice smaller than when encoded in the RDF/XML syntax.

The Protg plugin does not only generate the Manchester OWL syntax encoded ontology file, but it also creates a schema for the ontology ABox storage database, according to the contents of the ontology. The database schema and the ontology file are transferred to the mobile device. Together they customize the CAM middleware for a given domain.

The run-time CAM middleware architecture (right-hand side of Figure 5) consists of the following elements: the Manchester OWL Syntax parser, the Lightweight Ontology library (LOnt), a relational database, and the Context-Based Routing (CBR) protocol. The CAM middleware offers the following

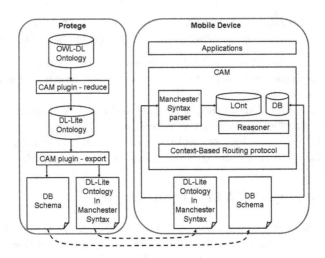

Fig. 5. Design-time and run-time CAM architecture

services to the applications: forming Context-Based Addresses, sending and receiving Context-Addressed Messages, and management of context information related to the node it executes on.

The Manchester OWL Syntax parser is generated automatically using the JavaCC [11] (Java Compiler Compiler) tool based on a Manchester OWL Syntax parser definition file for the WonderWeb OWL API [12].

LOnt is a library produced by the authors of this paper, capable of handling OWL-DL (not just DL-Lite) ontologies. It implements the interfaces of the Jena API [4] in a way suited for mobile devices. Its main feature is small size, and low memory requirements; it can successfully run on J2ME-enabled mobile phones. The CAM middleware uses LOnt for representing the TBox part of the domain model ontology.

The CAM middleware uses a relational database for storage and retrieval of the ABox of the ontology (i.e. context data) according to [5]. We have decided to use the HSQL [13] database for its performance on mobile devices.

Table 1. Comparison of Manchester Syntax and RDF/XML Syntax

Manchester OWL Syntax	RDF/XML Syntax
Class: FinishLine SubClassOf: StagePointOfInterest	<owl:Class rdf:ID="FinishLine"> <rdfs:subClassOf> <owl:Class rdf:about="#StagePointOfInterest" /> </rdfs:subClassOf> </owl:Class>

/**
* @rdfs:comment Assigns Mountain
* Passes to a specified Stage.
*/
Object property: hasMountainPass
    Domains: Stage
    Ranges: MountainPass
	<owl:ObjectProperty rdf:ID="hasMountainPass">
	<rdfs:domain rdf:resource="#Stage" />
	<rdfs:range rdf:resource="#MountainPass" />
	<rdfs:comment
	rdf:datatype="http:www.w3.org/2001/XMLSchema#string">
	Assigns Mountain Passes to a specified Stage.
	</rdfs:comment>
	</owl:ObjectProperty>

The Reasoner component implements the DL-Lite mechanism of conjunctive queries [5], on top of the LOnt library and the relational database. This conjunctive query reasoning algorithm is triggered whenever a new Context-Addressed Message is received by a node. The task of transporting sent messages between network nodes, is performed by the Context-Based Routing (CBR) protocol, which is described in a separate paper [6].

The usage of the CAM service can be summarized as follows:

1. Each node injects its local context into its ABox.
2. The application creates a Context-Based Address.
3. The application sends a Context-Addressed Message using the CAM service to the created address.
4. The CAM service sends the message to destination nodes using CBR.
5. CBR routes the message through the network.
6. Each node identified by CBR as a prospective destination checks if it actually is a destination for that message. This is done by invoking the Reasoner with the address of the message as a parameter. The Reasoner makes use of the TBox (domain model) stored in the LOnt library, and the ABox stored in the database (context data) to provide a response. We refer to this process as address resolving.
7. If the response of the Reasoner is positive, the message is passed from CBR to the CAM component and from there to the application layer.

Figure 6 presents a practical example of usage of the CAM service. The first part of the example presents how the context of a node can be defined. The second part shows how a Context-Based Address can be created, and how a message can be sent to that address. The third part shows how the sent message can be received at the destination. Each node in the network that defined its context as shown in the first part of the example will receive the message sent to the address created in the second part of the example by performing the receive operation shown in the third part of the example.

```
1) Store context
1. DomainInstance thisNode = cam.getThisNodeInstance();
2. thisNode.setInstanceOf(cam.getDomainClass(''Firefighter''));
3. thisNode.setInstanceOf(cam.getDomainClass(''Captain''));
4. DomainProperty p = cam.getDomainProperty(''isLocatedAt'');
5. thisNode.setObjectPropertyValue(p,''Jussieu Station'');
```
```
2) Send a CAM message
1. ContextBasedAddress a1,a2,a3,dest;
2. a1 = cam.makeCBA_InClass(''Firefighter'');
3. a2 = cam.makeCBA_HasInstanceValue(''isLocatedAt'',''Jussieu Station'');
4. a3 = cam.makeCBA_InClass(''Captain'');
5. dest = a1.makeIntersection(a2).makeIntersection(a3);
6. cam.sendMessage(dest,''Help!'');
```
```
3) Receive a message
1. CamConnection conn = cam.getConnection();
2. String msg = conn.waitForMessage();
3. System.out.println(''Message received''+msg);
```

Fig. 6. Usage of the CAM service

4   Performance Results

The test system is the Nokia N800 device[14], running the Linux based Internet Tablet OS 2007 Edition. The N800 offers a 320MHz processor and 128MB DDR RAM. The CAM middleware is written in Java and runs on top of Cacao JVM[15] using the HSQL[13] database. The test application invokes each tested method 100 times and calculates the average value of measured execution times. The ontology used for tests consists of 48 classes, 23 object properties and 31 datatype properties.

To evaluate the performance of the implemented Context-Addressable Messaging service we need to consider three mechanisms that affect overall performance in a major way: 1) domain model ontology (TBox) access, 2) context data (ABox) storage and retrieval, 3) reasoning.

Table 2 presents results measured for operations related to the domain model ontology (TBox) access, i.e., the performance of the LOnt library. We notice that the simple operations which just need to locate a specific resource and return some of its attributes are the fastest ones; however, the exact timing depends on how many relevant 'items' are there. For example we can see that the time measured for GET SUPER CLASSES operation is much shorter than for GET SUB CLASES operation, which is due to the fact that the class chosen for tests has only one super class, but a few sub classes. We can also observe the impact of reasoning, when we compare the 1,05ms obtained for GET SUB CLASSES method, with 15,78ms obtained for GET REASONED SUB CLASSES (the difference here is that the first method returns only direct sub classes of the selected class, while the latter iterates all the way down through the ontology).

Table 2. Domain model ontology access performance - LOnt

Operation	ms
GET PROPERTIES OF A CLASS	1,30
GET DIRECT SUB CLASSES OF A CLASS	1,05
GET DIRECT SUPER CLASSES OF A CLASS	0,65
GET REASONED SUB CLASSES OF A CLASS	15,78
GET REASONED SUPER CLASSES OF A CLASS	2,63
GET RANGE OF A PROPERTY	0,26
GET DOMAIN OF A PROPERTY	0,97
GET DIRECT SUB PROPERTIES OF A PROPERTY	0,12
GET DIRECT SUPER PROPERTIES OF A PROPERTY	0,12
GET REASONED SUB PROPERTIES OF A PROPERTY	0,08
GET REASONED SUPER PROPERTIES OF A PROPERTY	4,23

Table 3 presents results measured for operations related to context data (ABox) management, that is creating instances, assigning property values to those instances, getting these values, or removing the entered information. The main distinction from the methods presented in Table 2 is that these methods require access to the database which is used for storing instances, and do not make use of reasoning. The values presented in Table 3 depend mainly on speed of the underlying database, and can vary greatly depending on the chosen database and its configuration.

Table 3. Performance of selected context management operations for HSQL

Operation	ms
CREATE INSTANCE	11,36
REMOVE INSTANCE	16,65
SET INSTANCE OF	18,95
CHECK INSTANCE OF	5,10
UNSET INSTANCE OF	12,89
SET PROPERTY VALUE	12,08
GET PROPERTY VALUE	5,84
ADD PROPERTY VALUE	17,20
UNSET PROPERTY VALUE	12,94

Address resolving makes use of reasoning to evaluate the conjunctive query forming a Context-Based Address, by performing queries on the context storage database. Addresses used during tests are composed of atoms of the $B(x)$ type (as described in the first row of Figure 4). Figure 7 shows how the time required to perform address resolving depends on three different attributes of these Context-Based Addresses.

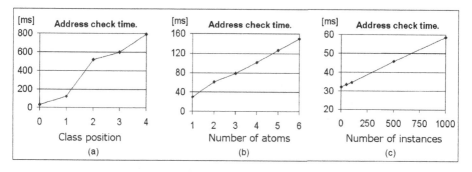

Fig. 7. Time required to resolve an address in the function of: (a) the position of class B in the taxonomy for a single class address, (b) the number of $B(x)$ atoms (where each class has no subclasses) in a DL-Lite query used as the address, (c) the number of other instances belonging to class B in the database for a single atom address composed of class B (which has no subclasses)

Figure 7a shows the time required to resolve an address in the function of class position in the taxonomy. The addresses used for this test are all composed of a single atom of the $B(x)$ type. The class position in the taxonomy is a value which indicates the level of class B in a taxonomy of classes (where level 0 indicates a class without subclasses). We observe that the time required to resolve an address which consists of a single class, increases with the position of that class in the taxonomy.

Figure 7b presents the time required to check an address in the function of the number of atoms in a DL-Lite conjunctive query used as a Context-Based Address. The addresses used for this test are all composed of $B(x)$ type of atoms, each with no subclasses. In this test we measured the time required to resolve an address which is a conjunction of N= <1,6> of such atoms.

Figure 7c shows how the time required to resolve an address depends on the amount of data in the ABox (context storage database). In this test we use addresses composed of a single $B(x)$ atom with no subclasses, and we create additional $N=<0,1000>$instances which belong to class B. We can observe that the time increases linearly with the number of other instances. However the influence of the number of other instance is relatively small compared to other factors.

We perceive the results achieved in the described configuration as satisfactory for lightweight traffic, such as e.g. a messaging service (similar to SMS) which is used for human generated traffic. For traffic intensive applications, such as machine to machine (M2M) communication, the delays introduced by the CAM middleware can be too large.

5   Related Work

The concept of semantic addressing of nodes applied in CAM, has been used in multiple research projects such as the EYES project [16], Content-Based Addressing [17] [18] or FlavourCast [19] (to name a few). However none of these projects aim to combine their addressing schemes with ontology based context modeling as done in CAM. To the best of our knowledge using expressions generated from an ontology as a Context-Based Address is a novel idea, which has not yet been researched, except for [20], where the authors extend the Siena subscription language by using ontologies and ontological operators. However this approach does not consider using expressions formed with multiple classes, and relies only on using single classes for an address.

Using ontologies as a model of context is in itself not a new idea (it has been taken into consideration in the context modeling survey [21] published in 2004). One of the proposed architectures for using an ontology based context model is an architecture where an ontology representation and reasoning is performed on a dedicated context interpreter machine which uses the JENA framework to resolve requests from mobile clients [22]. Other similar approaches exist, such as [23,24,25].

In the architecture presented in [26], ontologies are used as a model for context data obtained from sensors, and context synthesizing is performed by means of a naive Bayesian classifier in order to detect the type of situation in which the user is currently in.

The concept of handling ontologies on a mobile device has already been proposed in a few publications. For example [27] describes an architecture where reasoning is conducted on top of ontology data stored in a relational database, where a very large ontology is distributed between servers and mobile devices.

One of the key features of Context-Addressable Messaging is reasoning on mobile devices. To the best of our knowledge only a few reasoners have been developed for these environments, one of which is the KRHyper reasoner [28] - a software library based on J2ME.

6 Conclusions

This paper presents a mobile middleware for Context-Addressable Messaging. The middleware uses an ontology as the domain (context) model, as well as runtime-defined ontology-driven classes as Context-Based Addresses. It includes reasoning at prospective message recipients for Context-Based Addresses resolution. To achieve reasonable performance, the DL-Lite subset of OWL-DL is used as the ontology language. Major components, developed by the authors to implement the middleware, include a DL-Lite Protg plug-in, a lightweight, mobile implementation of the Jena API (the LOnt library), and a mobile DL-Lite reasoning engine. A Context-Based Routing protocol (CBR), used to limit the number of prospective recipients, is presented elsewhere [6]. The described Context-Addressable Messaging service offers satisfactory performance for human-to-human, SMS-like messaging.

Research problems to be addressed in the future include: (a) analysing the possibilities for automated ontology reduction from OWL-DL to DL-Lite, (b) introducing caching mechanisms to improve the performance of address resolving.

Acknowledgements

The authors would like to thank Ellen Munthe-Kaas from the University of Oslo, for a suggestion to use DL-Lite. We would also like to thank Matthew Horridge from the University of Manchester for help with the Manchester OWL Syntax parsing, and for sharing relevant source code with us.

References

1. Domaszewicz, J., Koziuk, M., Schoeneich, R.O.: A Context Addressable Messaging service for Mobile Ad-Hoc Networks. In: The 7th International Conference on Ontologies, DataBases, and Applications of Semantics - ODBASE 2008, Monterrey, Mexico, November 11 - 13 (accepted, 2008)
2. W3C Recommendation: OWL Web Ontology Language Guide. (10 February 2004), http://www.w3.org/TR/owl-guide/
3. Sirin, E., Parsia, B., Grau, B.C., Kalyanpur, A., Katz, Y.: Pellet: A practical owl-dl reasoner. Web Semant. 5(2), 51–53 (2007)
4. Carroll, J., Dickinson, I., Dollin, C., Reynolds, D., Seaborne, A., Wilkinson, K.: Jena: Implementing the semantic web recommendations (2003)
5. Calvanese, D., Giuseppe, D.G., Lembo, D., Lenzerini, M., Rosati, R.: DL-Lite: Tractable description logics for ontologies. In: Proceedings of the Twentieth National Conference on Artificial Intelligence, pp. 602–607 (2005)
6. Schoeneich, R.O., Domaszewicz, J., Koziuk, M.: Concept-Based Routing in Ad-Hoc Networks. In: The 10th International Conference on Distributed Computing and Networking - ICDCN 2009, January 3-6 (submitted, 2009)

7. Grau, B.C.: Owl 1.1 web ontology language, tractable fragments (19 December 2006), http://www.w3.org/Submission/owl11-tractable/

8. Knublauch, H., Fergerson, R.W., Noy, N.F., Musen, M.A.: The protégé owl plugin: An open development environment for semantic web applications. In: McIlraith, S.A., Plexousakis, D., van Harmelen, F. (eds.) ISWC 2004. LNCS, vol. 3298, pp. 229–243. Springer, Heidelberg (2004)

9. W3C Working Group: OWL Web Ontology Language Parsing OWL in RDF/XML (Note 21) (21 January 2004), http://www.w3.org/TR/owl-parsing/

10. Horridge, M., Drummond, N., Goodwin, J., Rector, A., Stevens, R., Wang, H.H.: The Manchester OWL Syntax. In: OWL: Experiences and Directions 2006 Athens, Georgia, USA, 10-11 November (2006)

11. Java Compiler Compiler [tm]: The java parser generator (2008), https://javacc.dev.java.net/

12. Bechhofer, S., Volz, R., Lord, P.: Cooking the semantic web with the owl api. In: Fensel, D., Sycara, K.P., Mylopoulos, J. (eds.) ISWC 2003. LNCS, vol. 2870, pp. 659–675. Springer, Heidelberg (2003)

13. HSQL: A java database engine v.1.7.3 (2005-02-07), http://hsqldb.org/

14. NOKIA N800: Product homepage (2008), http://www.nseries.com/products/n800/

15. CACAO: A java virtual machine (June 6, 2007), http://www.cacaojvm.org/

16. Handzinski, V., Koepke, A., Frank, Ch., Karl, H., Wolisz, A.: Semantic addressing for wireless sensor networks. Technical report, Telecommunication Networks Group, Technische Universität Berlin (2004)

17. Carzaniga, A., Rosenblum, D., Wolf, A.: Content-based addressing and routing: A general model and its application (2000)

18. Carzaniga, A., Wolf, A.L.: Forwarding in a content-based network. In: SIGCOMM 2003: Proceedings of the 2003 conference on Applications, technologies, architectures, and protocols for computer communications, pp. 163–174. ACM, New York (2003)

19. Cutting, D., Corbett, D.J., Quigley, A.: Context-based messaging for ad hoc networks, May 8-13 (2005)

20. Keeney, J., Lynch, D., Lewis, D., O'Sullivan, D.: On the role of ontological semantics in routing contextual knowledge in highly distributed autonomic system. Technical report, Department of Computer Science, Trinity College Dublin (2006)

21. Strang, T., Linnhoff-Popien, C.: A context modeling survey (2004)

22. Pessoa, R.M., Calvi, C.Z., Filho, J.G.P., de Farias, C.R.G., Neisse, R.: Semantic context reasoning using ontology based models. In: Dependable and Adaptable Networks and Services. LNCS, pp. 44–51. Springer, Heidelberg (2007)

23. Harry, C., Finin, T., Joshi, A.: An Intelligent Broker for Context-Aware Systems. In: Dey, A.K., Schmidt, A., McCarthy, J.F. (eds.) UbiComp 2003. LNCS, vol. 2864. Springer, Heidelberg (2003)

24. Dey, A.K., Abowd, G.D.: The context toolkit: Aiding the development of context-aware applications. In: Workshop on Software Engineering for Wearable and Pervasive Computing, Limerick, Ireland, June 6 (2000)

25. Gu, T., Wang, X., Pung, H., Zhang, D.: An ontology-based context model in intelligent environments. In: Communication Networks and Distributed Systems Modeling and Simulation Conference, San Diego, California, USA, (January 2004)

26. Korpipaa, P., Mantyjarvi, J., Kela, J., Keranen, H., Malm, E.-J.: Managing context information in mobile devices. IEEE Pervasive Computing 2(3), 42–51 (2003)
27. Specht, G., Weithoner, T.: Context-aware processing of ontologies in mobile environments. In: MDM 2006: Proceedings of the 7th International Conference on Mobile Data Management, Washington, DC, USA, p. 86. IEEE Computer Society, Los Alamitos (2006)
28. Kleemann, T., Siner, A.: Description logics based matchmaking on mobile devices. In: Workshop on Knowledge Engineering and Software Engineering, KESE (2005)

A Wearable, Conductive Textile Based User Interface for Hospital Ward Rounds Document Access

Jingyuan Cheng, David Bannach, Kurt Adamer, Thomas Bernreiter, and Paul Lukowicz

Abstract. This work is motivated by a hospital ward rounds scenario in the EU sponsored WearIT@Work Project. Based on a detailed application design and evaluation described in our previous work we have implemented a simple, wearable user interface seamlessly integrated in the doctor's coat. The interface is based on a multi-electrodes, conductive textile capacitive sensor that allows the doctor to control the system with simple gestures without the need to touch non sterile material. This paper focuses on the software that robustly extracts the gestures from noisy sensor signals and an extensive real life evaluation, including a two weeks' deployment of the interface in a hospital where it was used during real life rounds by three doctors, user questionnaires from a demonstration to a broader audience of hospital staff, and a systematic quantitative evaluation with students.

Keywords: capacitance measuring; wearable sensing.

1 Introduction

The work described in this paper is part of a larger effort, to optimize data access during hospital ward rounds sponsored by the EU WearIT@Work project [1]. It builds on a three year application design and evaluation phase described in [2]. A major result of this work has been the need for a simple input interface that allows the doctor to navigate through documents displayed at a bedside mounted screen.

Obviously the interface should not affect the doctor's ability to interact with patients or hinder him in performing examinations. This means that devices such as wrist worn keyboards and PDAs which require the use of both hands are unsuitable. Neither is voice control which would interfere with the conversation. An additional, non obvious requirement that limits the range of possible interfaces stems from the need to keep the doctor's hands sterile during the examination. In general the doctor will sterilize his hands before starting examining the patient. During the examination he should then avoid touching non sterile objects. Based on our talks with doctors, they are especially unwilling to touch electronic devices directly, which are difficult to sterilize and could potentially "accumulate" infections over time. The exact requirements of the scenario and their derivation are described in [2].

An initial solution has been to use a wrist mounted accelerometer for gesture recognition. However, this solutions has only had limited success. Person dependent training was not feasible, gestures had to be quite subtle not to irritate the patient,

D. Roggen et al. (Eds.): EuroSSC 2008, LNCS 5279, pp. 182–191, 2008.

there would be large variability in the sensor attachment and the doctors were not willing to invest time into learning how to operate the system. As a consequence, for most users the system did not work reliably enough while at the same time many still complained about the gestures being "too large". Again this has been described in detail in [2].

The above results have lead to the concept of a capacitive, conductive textile based sensor placed under the doctor's coat to detect small gestures performed in the proximity of the sensor location. The hardware design, the physical background and first step data analysis can be found in [3].

In this paper we describe the design and implementation of the actual interface and present the results of an extensive empirical evaluation. This includes a two weeks' deployment of the interface in a hospital where it was used during real life rounds by three doctors, user questionnaires from a demonstration to a broader audience of hospital staff, and a systematic quantitative evaluation with students.

1.1 Related Work and Paper Contributions

A similar concept to the one investigated in this paper was presented in [4]. It allows the user to swipe his hand across a watch that would detect the motions with an array of proximity sensors. The use of on body capacitive sensing for user interfaces has been proposed by [5]. Later the same group has used on body capacitive sensing for motion tracking in a dance application [6]. Capacitive gesture recognition for pervasive computing (but not wearable systems) has been discussed [7] (and a string of other publication by this group). In the wearable area capacitive sensing is also the basis of widely used textile pressure sensors [8].

Unlike the above examples our system has large area sensors (which mean that the doctor can operate it without having to pay attention to locating the right spot) and is made of textile so that it is flexible and can be easily integrated under the doctor's coat. Unlike the textile pressure sensor based interfaces it does not require touching or pressing and can be operated by just moving the hand slightly above the coat. The main contributions of the work described in this paper are:

a. Implementation of the interface software in such a way that it robustly recognizes the relevant gestures in real time and is not prone to false positives, despite week, noisy signals due to flexible, coat integrated interface.
b. Extended (two weeks) deployment of the entire system in a real life environment (during real word rounds with real doctors and patients)
c. Small, quantitative evaluation of the interface with real medical personnel
d. Large detailed quantitative evaluation with students.

2 System Overview

2.1 Overall Architecture and Functionality

The overall system concept is based on work described in detail in [2] (Fig. 1) and will only be briefly summarized below, as the focus of this paper is on the interface and system test. A bedside display is used to show patients' data from a hospital information system. When a doctor-nurse team arrives at the bed, they use wrist mounted RFID readers to read the tag on the patient's wrist. This identifies the patient, establishes the teams' authority to access his data, and brings up his record on the bedside display. The doctor can then browse the data using the textile touchpad (Control pad in Fig. 2) invisibly integrated (see next paragraph and [3]) inside of his coat. The nurse, who does not need free hands and can focus on the system, may access the record in parallel on a tablet PC and perform operations (such as ordering examinations) which are too complex for the doctors' interface.

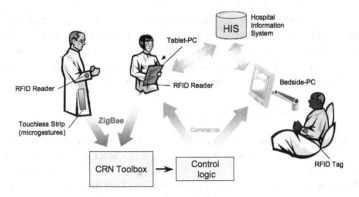

Fig. 1. Overall system architecture

2.2 Hardware Components

The three main components of the system (Fig. 2) are:

a. ID recognition: RFID-tags for doctor and patients, RFID-Readers for doctor and nurse
b. UI control: conductive textile touchpad under doctor's clothes controlled purely by hand
c. UI: wireless data communication and documents display

The ID recognition part is based on a SkyeTek RFID Reader M1-Mini, which is connected to a Moteiv Tmote-mini that sends data out via 2.4G ZigBee wireless connection. They are put together into a wristband and the RFID tags are also embedded into wristbands.

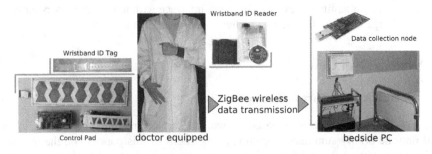

Fig. 2. System Components

The UI control part is made of a capacitive pad described in [3]. An ideal capacitor consists of two conductive plates separated and insulated from each other by a dielectric. Because human body is conductive, it can be considered as one of the electrodes. In our design, metal textile is fixed to the inner side of doctor's clothes with snap fastener and the doctor's hand is considered as the other electrode. Data are sent out via Tmote-mini, too. Two versions of pads were developed: a big one which is meant to be controlled by one hand, and a smaller one controlled by finger. Each pad is of 7 channels.

Both embedded boards (one for gesture control and one for RFID reading) have its own rechargeable battery and can keep running for more than 10 hours.

A Moteiv Tmote-sky is attached as data collection node to a bedside PC, on which data processing and user interface runs.

3 Software Design

3.1 General Structure

A hospital user interface called MPA has been developed by Systema under Windows for medical record management. The other part of software is divided into three parts and designed under CRN Toolbox [9]: data receiving and decoding, patients ID and gesture recognition, and UI command sending, as shown in Fig. 3. The main purpose of this part is to convert hardware level raw data to UI level information or commands.

Mapping RFID number to patient index can be achieved by an editable look-up table.

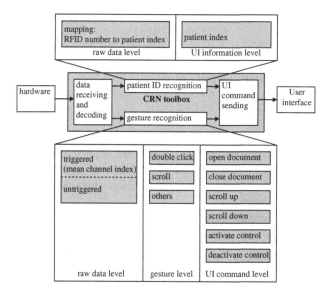

Fig. 3. Software structure under CRN Toolbox

As for gesture recognition, a gesture level is inserted between the raw data level and command level. Based on whether one or more channels are triggered and the mean channel index, raw data are grouped into three categories: double click, scrolling, and others, which are further used to switch finite state machine and send commands to user interface.

3.2   Raw Data Level Algorithm

Simply apply a fixed threshold to every channel will give out the triggered channels. However, for better performance, auto-adjust threshold are implied.

If channel i is triggered in the last data packet, $th(j,i)$, the threshold for channel j deduced from i, is:

$$th(j,i) = \frac{|j-i|}{6} \times (th_{high} - th_{low}) + th_{low} \tag{1}$$

Where th_{high} and th_{low} are the highest and lowest possible values of threshold.

$th(j)$, the final threshold for channel j is set to the minimum of all $th(j,i)$ s. Or if there is no channel triggered in the last packet, threshold for all the channels will be set to th_{high}.

The mean channel index is:

$$MCI = \frac{\sum\limits_{j,A(j)>th(j)} A(j) \times j}{\sum\limits_{j,A(j)>th(j)} A(j)} \tag{2}$$

Where $A(j)$ is amplitude of channel j.

3.3   Gesture Level Algorithm

Although theoretically this pad can provide more functions like single click, zoom in and out, finally only two groups of valid gestures are implied. Because the fewer gestures implied, the smaller the error rate is. In hospital scenario, where end users should not be expected as experts on computer, less gestures also means less effort in learning and using the system.

The mapping from raw data level to gestures is shown in Fig. 4.

Gesture recognition always starts with an un-triggered status. Double click is only triggered when the change of MCI is not too big. The time length for triggered and un-triggered status is also checked. The change threshold for MCI is set to 2 channels at the beginning, and reduces as the user scrolls further down. This provides more accurate scrolling in the beginning and quicker scrolling for long distance moving.

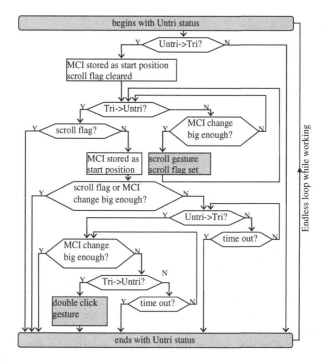

Fig. 4. Mapping raw data into gestures (MCI: mean channel index, Tri: triggered status, Untri: un-triggered status)

3.4 User Interface Level Algorithm

The mapping from gestures to user interface command is based on a finite state machine, as in Fig. 5. The program runs in two states: activated and deactivated. In deactivated status, it only responses to double click and then changes into activated status, where it reacts to double click as open/close(open when all documents closed, and close when a certain document opened), and scroll up/down. If no action was taken for a certain time, the program automatically returns to deactivated status.

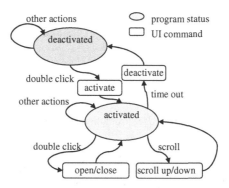

Fig. 5. Finite state machine for program states and UI commands

4   Evaluation

To improve the system, a test scenario was first set up in our lab. Then a two weeks real situation test was performed in Hospital Steyr, Austria from May 12th to 23rd 2008. After that, a quantitative evaluation was carried out during a demonstration in the hospital cafeteria. Finally, we repeated the student test again.

4.1   Test and Demonstration in Hospital Steyr

The system was set up in a double-beds patient-room with two bed-side PCs. The doctors used this system to view and show the patients' medical record twice a day, once in the morning and once in the afternoon. Each time it lasted for about 10 minutes. Altogether there were 3 doctors and 5 patients involved directly in the 10 days test. Doctors were satisfied with the system which was used continuously throughout the two weeks' period. Most patients suggested that the ability to go through the records together with the doctor was satisfying, and they were happy to see their own medical record on the screen. This test proved that the system could be used under real circumstance where situation was more complex with the involvement of non-technical persons and clinical devices.

However, due to law, privacy reasons, and the fact that we were constrained to three doctors it was impractical to perform any quantitative evaluation.

Hence a demonstration was held in the hospital's cafeteria afterwards, as shown in Fig. 6. We brought the whole system into cafeteria during lunch time, told people how to use it and then let them try. Finally 25 questionnaires were filled out after the two days' demonstration, among them 15 from doctors, 2 from nurses and 8 from others. The results were based on first impressions, because in general each person had only about 10 minutes to get to know and try the system.

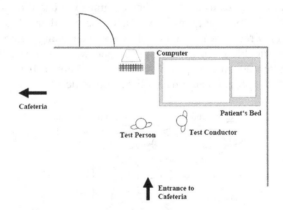

Fig. 6. End User Promotion in Cafeteria

Generally, people found the system worked well. There are 11 "no answer" because this question on general feeling was added only on the second day, when we received 14 valid feedbacks. When it comes to whether you would like to use such a system in daily work, 12 answered yes, 7 said no, and 6 remained undecided. The statistic details can be found in Table 1 and Table 2. We didn't include age, gender, status information in the cafeteria questionnaire, because people were supposed to be in a hurry so we put as less questions as possible.

Table 1. Statistics on general feeling about the system from cafeteria demonstration

	very much	much	not so much	only a little	not at all	no answer
reduce my stress on the job	1	4	6	7	3	4
increase my productivity	2	6	6	4	4	3
makes me uncomfortable	0	4	6	8	4	3
Increases the time I can spend with patients	2	5	7	6	2	3
affects my interaction with colleagues and/ or patients	0	9	6	3	2	5

Table 2. Statistics on function evaluation on the pad from cafeteria demonstration

	very good	good	not so good	bad	very bad	no answer
activate	8	16	1	0	0	0
scroll	6	7	11	0	0	1
open/close	8	14	3	0	0	0
in general	4	9	1	0	0	11

4.2 Student Test

As mentioned before, a student test has already been run once before the hospital test. But in order to put the system in its best mode for the coming hospital test, algorithms were changed constantly during the first student test when there were problems discovered or suggestions submitted. After the hospital test, the second round student test was carried out in our lab from May 27th to 30th 2008.

In the student test, a test person stood in front of a table, where a dummy with RFID-tag on its left wrist lay. The SmartBoard, a big display, was placed in the corner for both him and the "patient", as in Fig. 7.

Before the test, the test person was allowed first to try the system for about ten minutes, while the logger stood beside him, introducing the system, answering his questions and providing suggestions. After the subject got familiar with the control methods, he was required to carry out a whole test with the following steps:

[1] Greet the patient; get his ID number by the wristband RFID-reader
[2] Examine the patient's foot, open report document No. 2 then x-ray document No. 3.
[3] Examine the patient's hands and open the final report document
[4] Shut down the whole system by reading the ID-tag for the doctor himself, say goodbye to the patient

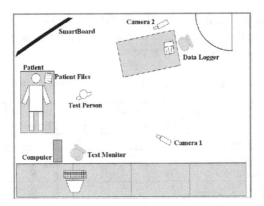

Fig. 7. Student tests in the university lab

Both big pad and small pad were tested. Altogether 10 persons joined the test, among them 7 computer science students, 2 teachers and 1 bank employee. In general, it takes less than twenty minutes to finish the tasks on both pads. People prefer the smaller pad than the big one, because it is easier to control with only one or two fingers than a complete hand, and also with smaller movement. Detailed statistics are shown in Table 3. Because the students are of the same generation and only 2 females took part in the test, we didn't make compare between age or gender group, as for the hospital cafeteria test.

Table 3. Statistic on function evaluation on the pad from student test

		very good	good	not so good	bad	very bad
big pad	activate	4	5	1	0	0
	scroll	0	2	7	1	0
	open/close	1	4	4	1	0
	in general	0	5	5	0	0
small pad	activate	6	3	1	0	0
	scroll	3	7	0	0	0
	open/close	3	6	1	0	0
	in general	3	7	0	0	0

5   Conclusion

The fact that it was used for two weeks in a real life hospital environment without major problems and to the satisfaction of both the doctors and the patients is a clear indication that our concept is well suited for ward round applications. Although only three doctors used the system during the deployment, the data from a more statistically relevant sample of medical staff (25 people) who tried the system during the cafeteria demo supports this assessment. The student tests further strengthen the argument and provide a more detailed insight into the positive and negative sides of the interface. The main issue to improve seems to be faster and smoother scrolling, which should be possible with a higher number of electrodes and faster circuits processing in the future.

Reference

1. Lukowicz, P., Timm-Giel, A., Lawo, M., Herzog, O.: WearIT@ work: Toward Real-World Industrial Wearable Computing. Pervasive Computing 6(4), 8–13 (2007)
2. Adamer, K., Bannach, D., Klug, T., Lukowicz, P., Sbodio, M.L., Tresman, M., Zinnen, A., Ziegert, T.: Developing a wearable assistant for hospital ward rounds: An experience report. In: Proc. of the Intl. Conf. on Internet of Things, pp. 289–307. Springer, Heidelberg (2008)
3. Cheng, J., Bannach, D., Lukowicz, P.: On Body Capacitive Sensing for a Simple Touchless User Interface, Hongkong. In: The 5th International Workshop on Wearable and Implantable Body Sensor Networks (BSN 2008), pp. 113–116 (2008)
4. Kim, J., He, J., Lyons, K., Starner, T.: The Gesture Watch: A Wireless Contact-free Gesture based Wrist Interface. In: 11th IEEE Intl. Symposium on Wearable Computers 2007, pp. 1–8 (2007)
5. Zimmerman, T.G., Smith, J.R., Paradiso, J.A., Allport, D., Gershenfeld, N.: Applying electric field sensing to human-computer interfaces. In: CHI 1995: Proceedings of the SIGCHI conference on Human factors in computing systems, pp. 280–287. ACM Press/Addison-Wesley Publishing Co., New York (1995)
6. Aylward, R., Paradiso, J.A.: Sensemble: a wireless, compact, multi-user sensor system for interactive dance. In: Proceedings of the 2006 conference on New interfaces for musical expression, pp. 134–139 (2006)
7. Wimmer, R., Holleis, P., Kranz, M., Schmidt, A.: Thracker-Using Capacitive Sensing for Gesture Recognition. In: Proceedings of the 26th IEEE International ConferenceWorkshops on Distributed Computing Systems, vol. 64 (2006)
8. Meyer, J., Lukowicz, P., Troster, G.: Textile Pressure Sensor for Muscle Activity and Motion Detection. In: 10th IEEE International Symposium on Wearable Computers 2006, pp. 69–72 (2006)
9. Bannach, D., Amft, O., Lukowicz, P.: Rapid Prototyping of Activity Recognition Applications. IEEE Pervasive Computing 7(2), 22–31 (2008)

Exploring the Design of Pay-Per-Use Objects in the Construction Domain

Daniel Fitton, Vasughi Sundramoorthy, Gerd Kortuem, James Brown,
Christos Efstratiou, Joe Finney, and Nigel Davies

Computing Department, InfoLab21, South Drive, Lancaster University, Lancaster, UK
{df,v.sundramoorthy,kortuem,jb,efstrati,joe,
nigel}@comp.lancs.ac.uk

Abstract. Equipment used in the construction domain is often hired in order to reduce cost and maintenance overhead. The cost of hire is dependent on the time period involved and does not take into account the actual use equipment has received. This paper presents our initial investigation into how physical objects augmented with sensing and communication technologies can measure use in order to enable new pay-per-use payment models for equipment hire. We also explore user interaction with pay-per-use objects via mobile devices. The user interactions that take place within our prototype scenario range from simple information access to transactions involving multiple users. This paper presents the design, implementation and evaluation of a prototype pay-per-use system motivated by a real world equipment hire scenario. We also provide insights into the various challenges introduced by supporting a pay-per-use model, including data storage and data security in addition to user interaction issues.

Keywords: Smart Objects, Pay-Per-Use, User Interaction, Mobile HCI.

1 Introduction

The application of pay-per-use payment models is receiving growing interest and is perhaps exemplified by numerous existing road pricing systems deployed around the world and various examples of pay-per-use car insurance [1][2]. Pay-per-use or 'metered service' is a payment model that allows customers unlimited access to services (such as electricity, phone calls, internet access etc.) or physical objects where cost is calculated according to usage. Many new examples of pay-per-use systems in use today have been enabled by advances in technology and motivated by economic factors (in the case of car insurance) or social factors (in the case of congestion related road pricing). While many examples of 'metered service' already exist in our daily lives the application of a pay-per-use model in the broader context of everyday objects has received very little attention. Pay-per-use can potentially be applied to any object into which a system to record usage can be embedded. To date, the only example of a pay-per-use object intended to investigate this idea is a prototype chair that contains sensors and can record use [3].

D. Roggen et al. (Eds.): EuroSSC 2008, LNCS 5279, pp. 192–205, 2008.

We explore this new research area by applying the pay-per-use concept to an equipment hire scenario in the construction domain. When construction companies deal with short-term contracts, it is common for equipment to be hired for the duration of that contract (in order to avoid purchase cost and maintenance overhead). However, the hire cost is currently proportional to the time period involved so regardless of whether equipment is used heavily or not at all the construction company will be charged the same amount. Additionally, those responsible for managing the hire of equipment have no accurate knowledge of how much use equipment is receiving and whether project management decisions need to be revised, for example, in order to reduce cost or increase productivity.

Our approach to addressing these problems is to augment individual pieces of equipmentx with sensing, storage and communication capability to enable them to become 'smart' objects. This work builds upon existing research using smart physical objects for safety monitoring on construction sites [4] [5] and we use smart objects to detect and store *experiences*. We use the term *experience* to refer to events or activities involving the object which it can detect. Detection of usage experiences in particular is then used to implement a pay-per-use billing model for equipment hire.

Smart physical objects not only allows the implementation of services such as pay-per-use but also provides the opportunity to enable a range of novel user interaction experiences. A piece of equipment used in the construction domain is involved in a range of meaningful experiences each connected with a variety of contextual information. Our goal is to make this information accessible to an interested (and authorised) party interacting via a mobile device. For example, a supervisor finding a discarded piece of equipment at a site could interact with it to find information such as whether it is in working order and how much its hire has cost to date.

Section 2 provides an overview then analysis of the construction scenario which motivates this work, followed by discussion of our pay-per-use billing model. Section 3 discusses the design of the pay-per-use smart object architecture. Section 4 provides an overview of the hardware and software implementation of the prototype system. Section 5 provides evaluation and discussion of the prototype including areas for future work. Section 6 discusses related work and section 7 presents concluding remarks.

2   Pay-Per-Use for Construction Equipment

2.1 Construction Scenario: Road Patching

The design of our pay-per-use prototype is motivated by the scenario of road patching (where defects to tarmac road surface are repaired). The task of road patching begins with a contract between a governmental body such as a local council and a construction company. The local council then provides details of the roads that need to be repaired. Typically, once the contract for work is in place, the construction company (the lessee) hires necessary equipments from a hire company (the lessor) for the duration of the contract. Figure 1 shows a site where road patching is taking place and hired the equipment in use.

Fig. 1. Key Equipment Used for Road Patching: 1 – Van, 2 – Petrol Powered Drum Roller/Air Compressor (Stored on Trailer), 3 – Air-powered Pavement Breaker, 4 – Petrol Powered 'Wacker' Plate Compactor

2.2   Analysis of Usage Scenario

Figure 2 depicts the main stages in the hire operation of a piece of equipment in a construction scenario. Within Figure 2 a range of users are involved, each with different responsibilities and goals. We divide users into two main categories. Firstly, employees from a company that lease equipment (stages 1, 1b, 2, 3 and 7). Secondly, we consider employees from a construction company that hires and utilises equipment with responsibilities summarised as:

- Administrator: Procuring and managing the hire of equipment (stages 2, 3 and 6),
- Supervisor: Managing and overseeing the deployment of workers and equipment (stages 4 and 5),
- Worker: Transporting equipment to a site and utilising it to carry out road patching (stages 5 and 6).

For the design of a smart object in a this scenario, three main areas to address can be identified; experiences the smart object must detect, user interactions that take place with the smart object and processes with which the object must take part. One example of each is discussed in the following paragraphs:

Usage Experience - The key aspect of the scenario for pay-per-use is the detection of usage experiences once workers have transported equipment to the required site. The equipment continually senses its state and stores usage experiences. If a piece of equipment develops a fault (which can not be sensed internally by the tool) a supervisor would interact to record this information. If an exception condition is sensed during usage (such as overheating, exceeding duty cycle, dropping/subjecting to excessive forces etc.) this mishandling experience is also recorded.

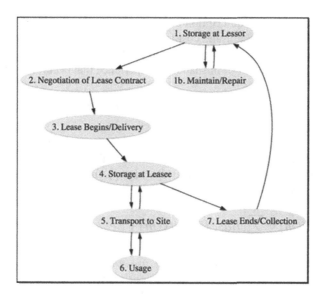

Fig. 2. Overview of Stages in Hire Operation of a Single Piece of Equipment

Query Interaction - This occurs when a supervisor visiting a remote site needs to view how much use a piece of equipment has received and the associated cost. The supervisor locates the equipment (which may be on a van or in the field) then interacts. In addition to cost, the supervisor is also able to view daily breakdowns of use (including per-worker) and any faults that have been reported.

Delivery Process - This concerns delivery of equipment by a hire company to a construction company depot after negotiation of a lease contract. At the point of delivery an employee of the hire company (lessor) interacts to confirm information relevant to the hire is stored on the device (date of delivery, cost information, point of contact at hire company etc.) and ensures that usage counters are reset. An administrator from the construction company (lessee) interacts to check equipment fulfils the required technical specification and confirm that delivery has taken place.

2.3 Pay-Per-Use Billing Model

A wide range of possibilities exist for calculating cost from usage. For example, charges can vary based on times of use (with peak and off-peak periods), usage level (such as cost reducing and usage increases) or even the intensity of use. We use a billing model which considers an object's usage and mishandling experiences. Equipment mishandling includes overheating, exceeding duty cycle, dropped/subjecting to excessive forces etc. We assume that if a piece of equipment is mishandled by the lessee additional maintenance (carried out by the lessor) is required and therefore a charge should be levied.

We now give the formal description of our own pay-per-use billing model. *Lease-Cost* is the summation of the usage cost (where T is the total period of use and U is the equipment-specific cost per usage unit) and total costs for every type of mishandling experience (*M* is the set of mishandling experience that occur, m ∈ M):

$$LeaseCost = (T \times U) + \sum_{i=1}^{|M|} Cost(m)_i \qquad (1)$$

The function *Cost(m)* returns the cost for each type of mishandling event *m*, for the number of times the event occurs, N; this cost sums the base penalty for the occurrence of the event (returned by the *BasePenalty(m)* function) with a severity surcharge (returned by the *Severity(m)* function):

$$Cost(m) = \sum_{n=1}^{N} [BasePenalty(m) + Severity(m)]_n \qquad (2)$$

In order to capture pay-per-use data we use wireless sensor nodes attached to the exterior of equipment. Equipment used in our construction scenario (as shown in Figure 1) exhibits vibration between certain ranges when in use (dependent on the individual piece of equipment). The sensor nodes contain an accelerometer and this is used to detect vibrations that indicate usage. The accelerometer is also used to detect experiences that involve sudden motions such as dropping (and the severity of the drop). The sensor nodes can potentially be used to detect other parameters in order to infer experiences such as monitoring temperature in order to detect overheating experiences.

3 Pay-Per-Use Smart Object Architecture

As discussed in the previous section, our approach is to attach wireless sensor nodes to equipment which detect experiences such as usage. The self-contained sensor nodes maintain a semi-persistent memory of their experiences without reliance on other nodes or additional infrastructure. This independence is necessary to simplify deployment and ensure reliability in the context of remote building sites with limited connectivity and ad-hoc movement and usage of equipment. However, a backend database for persistent storage of experiences does exist and the use of mobile devices to interact with objects provides a form of opportunistic connectivity for transferring data. We assume that all users which interact with the smart objects have personal mobile devices such as mobile phones or PDAs, but these may have varying communication, storage and user interaction capabilities.

The overall architecture of the system is shown in Figure 3, where each piece of equipment is augmented with a wireless sensor node (labelled Asset Tag) which provides sensing, storage and processing. Mobile devices then interact with a sensor node (shown by the black arrow) and enable data to be transferred to a backend infrastructure (the dashed areas indicate future work). In summary, the sensor node interprets

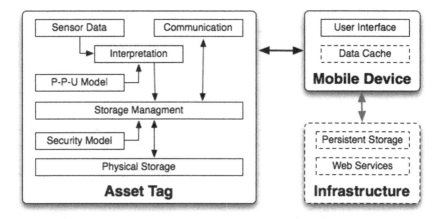

Fig. 3. Architecture of Prototype System Implementation

low-level sensor data to detect higher-level experiences of interest (e.g. equipment in use). Before the experiences are committed to physical storage, security and pay-per-use models are applied. Additionally, use of physical storage is carefully managed to ensure optimum use of space available and flash RAM (with a finite number of reads and writes). The mobile device is allowed to access data stored on the sensor node, subject to the security model. It is our intention that each user should be presented with their own personalised user interface based on the information they can access, the tasks which they are able to carry out and the capabilities of their mobile device. We now discuss the design of the security, communication and pay-per-use model aspects of the architecture in more detail.

While the key requirement for implementing the pay-per-use prototype is the detection and storage of usage experiences, we also wish to record involvement in processes that cannot be sensed (such as delivery) about which equipment has to be 'told' through user interaction. Storing a wider variety of information on a piece of equipment that is passed among different users in different companies raises the need for a security model. For example, the lessor is unlikely to allow the lessee to access information related to the profit generated though the lease of a piece of equipment. Conversely, the lessee may have to carefully control access information pertaining to individual workers use of the equipment.

Information stored at a piece of equipment may have been authored by the equipment itself (for example, interpreted from sensor readings), the lessor or the lessee. In order to provide data security, the categories of users or specific user that can access information are explicitly specified. In Table 1 we classify some example pieces of information which are stored on a sensor node in our prototype system according to the author and security restrictions.

Devices attached to equipment in the field have potential for only limited, opportunistic network connectivity with external back-end infrastructure. Therefore, embedded storage is essential for recording and accessing experiences that occur in real time. This requirement presents additional challenges for managing data on a device with limited storage capacity. However, the mobile phone used for interaction may also transfer data from the equipment to a backend infrastructure via network technologies such as GPRS/UMTS and also when synchronising the mobile device with a desktop machine.

Table 1. Examples of Data Stored on the Pay-Per-Use Prototype

Author	Information	User Access
Sensor Node	Total use for current lessee	Lessor, Lessee: Administrator, Supervisor
	Per-worker usage for current lessee	Lessee: Supervisor, Worker (to which the usage is attributed)
	Maintenance experiences	Public
	Exception experiences for current lessee	Public
Lessor	Contact details	Public
	Pay-Per-Use billing model details	Lessor
	Serviceable lifetime of equipment remaining	Lessor
Lessee	Lease contract details	Lessee: Administrator, Supervisor
	Workers assigned to equipment use	Lessee: Supervisor, Worker

As discussed in section 2.4, a wide variety of pay-per-use billing calculations are possible which, in addition to usage time, could taking into account other parameters such as usage intensity and usage location. In our prototype scenario we currently implement the billing calculation at the piece of equipment in order to ensure accurate cost information is always available when interacting. However, we also assume that verifiable usage data retrieved during user interaction during and/or at the end of lease contract is used by the lessor to generate formal invoices.

4   Prototype Implementation

4.1   Hardware

The current prototype implementation is based on an existing hardware platform (Nemo Tags [6]) with wireless sensor nodes that include an accelerometer, ARM processor and low-power 802.15.4 radio enabling communication with and proximity detection. The platform is designed to enable HAV monitoring for extended periods in the field and consists of an Asset Tag which is fixed to equipment. Asset Tags are housed in a rugged enclosure (as shown in Figure 5) to deal with the harsh environment of building sites. They remain continually powered on (power management is a key consideration [6] [7]) and automatically detect use without any user intervention.

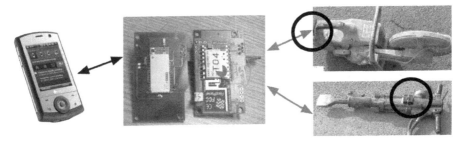

Fig. 5. Overall Architecture Showing Mobile Phone Interaction with Equipment (Nemo Asset Tags Circled)

In order to enable mobile phone interaction with the Nemo Asset Tags a Bluetooth extension board was developed in order to provide a Bluetooth Bridge Tag. This enables communication between Nemo Asset Tags using 802.15.4 and Bluetooth enabled mobile devices. As interaction with equipment will take place in the harsh environment of construction sites, it was important to support a mobile device that was available in ruggedized format. Only a limited range of ruggedized phones that support development of 3rd party applications are available and as the majority of these are Windows Mobile devices this platform was chosen. Figure 5 shows the overall architecture for the prototype scenario where the black arrow represents Bluetooth communication and the grey arrow 802.15.4.

4.2 Software

The applications that run on the Bluetooth Bridge Tag and Asset Tags attached to equipment are developed in C. The Bridge Tag continually listens for commands sent by a phone, validates those received and sends messages over the 802.15.4 radio as appropriate. A simple textual command set is currently used for all communication. Data received over the 802.15.4 radio is checked for integrity and forwarded to the phone if appropriate.

The mobile phone application, currently implemented in C#, initially discovers Bluetooth devices in range and allows the user to connect to the desired Bridge Tag. Once paired, communication takes place using Bluetooth SPP (Serial Port Profile) and commands are sent between the phone and bridge. An Asset Tag discovery processes is initiated from the Bridge Tag (broadcast over the 802.15.4 radio) and tags in range reply with basic information such as equipment name, leaser name and status. The results of the discovery are used to populate a GUI on the phone application (Figure 6). Once the user has identified a piece of equipment they wish to interact with, they then 'connect' to it and access information via a customised user interface (currently implemented with a different application for each group of users). The interface for a supervisor in a construction company is show in Figure 7.

This early prototype implementation of the application and user interfaces is based on our analysis of the scenario and technical feasibility of features. We plan to refine our design through user studies with workers involved with equipment hire and use in the construction domain.

Fig. 6. Phone Interface Showing Results of Discovery Process

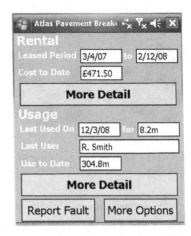

Fig. 7. Construction Company Supervisor User Interface

5   Evaluation and Discussion

5.1   Quantitative Evaluation of Storage Model

A fundamental requirement for pay-per-use implementation is persistent storage of usage experiences. Fulfilling this requirement in a resource-constrained embedded implementation requires careful management of available storage. From our analysis of usage data generated by a pavement breaker we found it possible to optimize the storage model.

From observations made during a field trial, typical use of a pavement breaker involves approximately 10 short bursts of drilling during a minute of use. The pavement breaker may be in continual use for upwards of 10 minutes in total (the period). Rather than logging raw data from the accelerometer, every continual burst of use is detected as a single usage experience. Assuming the record structure shown in Table 2 is used to store usage experiences where 20 bytes of storage are required to store every individual burst of drilling. The existing Nemo Asset Tag hardware implementation provides 460,800 bytes of available flash memory, which can potentially hold a maximum of 2304 minutes or 38 hours of use (assuming no other data is stored).

This storage limit may be exceeded when considering a piece of equipment that may be used heavily in the field over a lease period of several years. The solution we have to this problem is to apply a form of simple data aggregation (also called data fusion) [8] to aggregate the individual usage experiences during the overall period of drilling (10 minutes or more) into a single experience. This yields a minimum hundred-fold increase in storage capacity. Aggregating the data is effectively trading-off detail in the resolution of the data collected for overall storage capacity. However, the trade-off here is acceptable as the finest granularity of information we wish to capture is individual periods of drilling.

Table 2. A Record Structure for Storing Usage Events

User (8 bytes)	Time Start (8 bytes)	Use Duration (4 bytes)
Unique User Tag ID	Seconds since epoch	Total in seconds

5.2 Evaluating the Trace of a Pay-Per-Use Example

We now consider the trace of experiences generated during a pay-per-use rental example and demonstrate the calculation of total cost for the customer. We consider the use of a pavement breaker over the course of five days with usage most days (shown in Table 3) and a single mishandling event (shown in Table 4). We detect use by comparing readings from the onboard accelerometer with known values which indicate use (recorded during testing in field trials), mishandling events such as dropping are detected in a similar manner.

Table 3. Usage Experience Records

User	Time Start	Use Duration
9458001000000007	1212312600	429
9458001000000007	1212318907	365
9458001000000009	1212329054	671
9458001000000032	1212487974	967
9458001000000007	1212657043	243
9458001000000007	1212674247	401

Table 4. Mishandling Experience Record

Time	Type	Severity Level
1212695901	1	12

Using the calculations presented in 2.4 we now find the total cost for the usage represented in tables 2 and 4. Using a per-second usage cost of £0.02, the total period of use in Table 3 (3076 seconds) the total usage cost is £74.12. Table 4 shows a single mishandling event, assuming *BasePenalty(1)* returns a value of £3 and *Severity(12)* returns a value of £5 the total mishandling charge is £8. The total cost for the lease period is £82.12 and the total amount of storage required for the records shown in Tables 3 and 4 is 134 bytes. This realistic example highlights the feasibility of our pay-per-use design and is easily accommodated by our prototype system.

5.3 Discussion and Future Work

Several issues remain to be addressed in the prototype implementation. In order to provide security in a scenario with groups of users with different privileges to access data (lessor, lessee etc.) we currently intend to implement this using encryption. The

802.14.5 radio module used on the Asset Tags [9] contains a hardware implementation of AES (Advanced Encryption Standard, also known as Rijndael) which enables the encryption of data in hardware during transmission, potentially removing the need to encrypt data stored on an Asset Tag. An important aspect of accurate pay-per-use billing is maintaining the integrity of usage data collected and stored on an object and this could be achieved though additional encryption (or even through the use of digital signatures and a trusted third party). Moving data from a pay-per-use object in the field to a back-end infrastructure (without resorting to the deployment of additional communication infrastructure such as that described [10]) is an area we wish to explore in the future. We consider that opportunistic connectivity provided by user interactions with objects from mobile phones is an excellent opportunity for this. However, challenges are raised if an object in the field cannot communicate with a back-end infrastructure directly and data has to travel via multiple asynchronous 'hops'. For example, an object would require confirmation that a piece of data has been successfully transferred to a back-end database intact before removing it from its own storage.

The use of a Windows Mobile device enabled rapid prototyping of user interfaces and interaction with the Nemo Asset Tags, in addition to the potential provision of a ruggedized user device. However, these devices require user interaction via a small keypad or keyboard and touch screen using a stylus. The use of these interaction methods, while not exclusive to Windows Mobile devices, require a very high degree of attention on the users part which may prove problematic on a building site where a user must be continually aware of their surroundings to avoid injury. A user may also be required to wear safety equipment on a construction site such as eye, ear and hand protection which may further increase the difficulty of interacting with a mobile device. It may be possible to help mitigate these problems through HCI design but an important aspect of future work will be discussion with industrial partners in order to explore these pragmatic user interaction challenges.

Refining and expanding the design of functionality provided by the prototype is a broader goal for future work. We plan to consider this extension in terms of what is technically feasible, what is desirable by users and how to support adoption of the system as part of users existing working patterns. We wish to explore these requirements through discussion of current and potential features with industrial partners using design workshops and other user-centred techniques.

6 Related Work

The idea of pay-per-use is traditionally applied to utility services (gas, electricity, phone, broadband etc.) but has recently been used in more dynamic automotive applications such as road pricing, where charges levied for the use of roads are also intended to reduce traffic congestion (the key concepts and examples of road pricing system are discussed in [11]). The concept has also been applied to car insurance, such as Norwich Union's 'pay-as-you-drive' product [1], where the customer attaches a device to their car that records use of the car and sends it to the insurer without user intervention. The Smart Tachograph [12] is a prototype system which takes this idea further and combines both road pricing and pay-per-use insurance. Smart

Tachograph is similar to the work presented here in that it exploits the technology in use to provide more than purely pay-per-use functionality (such as providing a user interface). When inferring collecting car usage information in the form of GPS data (used in the 'pay-as-you-drive' product and Smart Tachograph) privacy issues emerge when disclosing data. GPS data can be analysed to determine factors such as whether the car exceeded speed limits and severity of acceleration/braking: information which can be used to determine whether a driver presents a high risk and should be charged more for their insurance. This potentially negative aspect of pay-per-use for certain users may emerge through our work. For example, through implementing pay-per-use some construction companies may find that they are paying more for hire of equipment than with a previous static scheme even though the extra charge is justified.

Related work describing existing prototypes of pay-per-use applied to objects is currently limited to a brief discussion of a pay-per-use chair idea [3]. However, the technological and financial motivations for implementing the prototype is similar to the work presented here. Several examples of prototypes that enable interactions with the environment via mobile devices exist, many use RFID tags and phones with in-built RFID tag readers. For example, the PERCI (PERvasive serviCe Interaction) architecture [13] enables interaction between the user and a set of semantic web services which compose the interactive aspects of the environment. In common with the aims of this work, the PERCI architecture supports the generation of user interfaces based on the capabilities of the user's mobile device. However, a phone equipped with an RFID tag reader is required and continual communication with web services using GPRS. Existing research has investigated a range of techniques for using mobile devices to interact with the environment which include touching, pointing, scanning and user mediated [14]. The technique of 'touching' is exemplified by RFID where a user has to physically touch an object to interact and 'scanning' describes interaction mechanisms such as Bluetooth where a user has to scan for objects in their current proximity. Pointing is typically implemented using a camera phone which recognises graphical markers located on an object and user mediated interaction involves a users specifying the object with which they wish to interact using an identifier such as a URL. Scanning is currently the most suitable method of interaction in our scenario as the user will be in close proximity of an object but physical contact may not be possible.

There exists extensive research involving wireless sensor networks (WSNs) in a variety of areas [15] [16]. Where large volumes of data are generated at nodes or support for queries is required often a distributed query processing system such as tinyDB [17] is used. However, the application of this research is largely focussed on utilising data collected from communication between numerous (highly resource constrained) nodes to monitor an environment. While the application of sensor nodes may differ in WSNs several examples of hardware platform used (such as BTnodes [18]) are similar to that used in this work.

7 Conclusion

This paper has presented our initial exploration of the design, implementation and evaluation of a prototypical system for enabling pay-per-use. The approach used was to augment physical objects with wireless sensor nodes to enable them to become smart objects. The scenario used to motivate the design of our prototype is based on equipment rented for use in the construction domain. Once augmented with the prototype sensor nodes, equipment detects and collects its own experiences (such as usage and mishandling) then applies a pay-per-use billing model to calculate cost. We also support interaction with the smart objects via mobile devices from a range of users involved the equipment hire and usage scenario. These interactions are not restricted to pay-per-use applications, for example an administrator responsible for equipment procurement may interact to view usage data in order to revise hire decisions and a site supervisor may interact to report a fault with the piece of equipment. While the design and implementation of the prototype is at an early stage, a range of technical challenges have been identified and considered. These challenges included supporting user interaction with a smart object, data security issues in the context of multiple users and the need for managing the storage of experiences in an independent resource constrained sensor node. In addition to further refinement of design and evaluation of user interfaces, several additional issues were identified for future work. These included enabling user interaction with mobile device in an environment where a user must be continually aware of their surroundings to avoid injury, and transferring data to a backend infrastructure using opportunistic connectivity via mobile devices. Pay-per-use systems such as our prototype enable a wide variety of new and novel applications and we hope the work presented here will be valuable to other researchers exploring this area.

References

1. Norwich Union Pay As You Drive Insurance,
   http://www.norwichunion.com/pay-as-you-drive
2. MyRate Program,
   http://auto.progressive.com/progressive-car-insurance/
   myrate-default.aspx
3. Accenture Technology Labs, Pay-per-use Object,
   http://www.accenture.com/Global/Services/
   Accenture_Technoogy_Labs/R_and_I/PayPerUseObject.htm
4. Kortuem, G., Alford, D., Ball, L., Busby, J., Davies, N., Efstratiou, C., Finney, J., White, M.I., Kinder, K.: Sensor Networks or Smart Artifacts? An Exploration of Organizational Issues of An Industrial Health and Safety Monitoring System. In: Krumm, J., Abowd, G.D., Seneviratne, A., Strang, T. (eds.) UbiComp 2007. LNCS, vol. 4717, pp. 465–482. Springer, Heidelberg (2007)
5. Davies, N., Efstratiou, C., Finney, J., Hooper, R., Kortuem, G., Lowton, M.: Sensing Danger – Challenges in Supporting Health and Safety Compliance in the Field. In: 8th IEEE-Workshop on Mobile Computing Systems and Applications (HotMobile 2007), Tucson, Arizona, February 26-27 (2007)
6. Efstratiou, C., Davies, N., Kortuem, G., Finney, J., Hooper, R., Lowton, M.: Experiences of Designing and Deploying Intellignent Sensor Nodes to Monitor Hand-Arm Vibrations in the Field. In: ACM MobiSys 2007, San Juan, Puerto Rico, pp. 127–138. ACM Press, New York (2007)

7. Brown, J., Finney, F., Efstratiou, C., Green, B., Davies, N., Lowton, M., Kortuem, G.: Network Interrupts: Supporting Delay Sensitive Applications in Low Power Wireless Control Networks. In: Proceedings ACM MobiCom 2007 Workshop on Challenged Networks (CHANTS 2007), Montreal, Canada, September 14 (2007)
8. Heinzelman, W.R., Chandrakasan, A., Balakrishnan, H.: Energy-Efficient Communication Protocol for Wireless Microsensor Networks. In: 33rd Hawaii international Conference on System Sciences, vol. 8. IEEE Press, Washington (2000)
9. Chipcon, CC2420 datasheet (2004),
   `http://www.chipcon.com/files/CC2420_Data_Sheet_1_3.pdf`
10. Kirisci, P.T., Hunecke, H.-H., Hribernik, K.A., Dikici, C.: A wireless solution for mobile collaboration on construction sites. In: IEEE International Workshop on Wireless Ad-Hoc Networks, Oulu, Finland, pp. 166–171. IEEE Press, Los Alamitos (2004)
11. Kelly, F.: Road pricing. Ingenia 29, pp. 34–40 (2006),
    `http://www.ingenia.org.uk/ingenia/issues/issue29/kelly.pdf`
12. Coroama, V.: The Smart Tachograph - Individual Accounting of Traffic Costs and Its Implications. In: Fishkin, K.P., Schiele, B., Nixon, P., Quigley, A. (eds.) PERVASIVE 2006. LNCS, vol. 3968, pp. 135–152. Springer, Heidelberg (2006)
13. Rukzio, E., Paolucci, M., Wagner, M., Berndt, H., Hamard, J., Schmidt, A.: Mobile service interaction with the web of things. In: 13th International Conference on Telecommunications, Funchal, Madeira island, Portugal, May 9-12 (2006)
14. Rukzio, E., Broll, G., Leichtenstern, K., Schmidt, A.: Mobile interaction with the real world: An evaluation and comparison of physical mobile interaction techniques. In: European Conference on Ambient Intelligence, Darmstadt, Germany, November 7-10 (2007)
15. Akyildiz, I.F., Su, W., Sankarasubramanisam, Y., Cayirci, E.: Wireless sensor networks: a survey. Computer Networks 38(4), 393–422 (2002)
16. Romer, K., Mattern, F.: The Design Space of Wireless Sensor Networks. In: IEEE Wireless Communications, pp. 54–61 (2004)
17. Madden, S.R., Franklin, M.J., Hellerstein, J.M., Hong, W.: The design of an acquisitional query processor for sensor networks. In: 22nd ACM SIGMOD International Conference on Management of Data, pp. 491–502 (2003)
18. Beutel, J., Kasten, O., Mattern, F., Romer, K., Siegemund, F., Thiele, L.: Prototyping wireless sensor network applications with BTnodes. In: Karl, H., Wolisz, A., Willig, A. (eds.) EWSN 2004. LNCS, vol. 2920, pp. 323–338. Springer, Heidelberg (2004)

Synthesizing Context for a Sports Domain on a Mobile Device

Alisa Devlic[1,2], Michal Koziuk[3], and Wybe Horsman[4]

[1] Appear Networks, Kista Science Tower,
16451 Kista, Sweden
alisa.devlic@appearnetworks.com
[2] Royal Institute of Technology (KTH), Department of Communication Systems,
Electrum 418, SE-164 40 Kista, Sweden
devlic@kth.se
[3] Institute of Telecommunications, Warsaw University of Technology,
ul. Nowowiejska 15/19, 00-665 Warsaw, Poland
mkoziuk@tele.pw.edu.pl
[4] Capgemini NL bv, Papendorpseweg 100
3528BJ Utrecht, The Netherlands
wybe.horsman@capgemini.com

Abstract. In ubiquitous computing environments there are an increasing number and variety of devices that can generate context data. The challenge is to timely acquire, process, and deliver these data to context-aware applications. The role of context synthesis is to generate new knowledge, as a result of a reasoning process applied to context information that is already present in the system. The success of this mechanism mainly depends on the response time that the end-user or an application must wait for the response to a context query. This paper describes and evaluates an approach to context synthesis on a mobile device to be used by a set of applications in a sports domain. A scenario based on a live race at the Super Prestige Cyclocross in Gieten, Netherlands demonstrates the use of context synthesis to dynamically compose gaps and groups of cyclists in order to provide a nearly real-time virtual ranking service.

Keywords: context synthesis, context operators, context modeling, sport scenario.

1 Introduction

Imagine experiencing a sport event from your phone, where you are your own director deciding upon your own point of view by actually moving about the event locale. Rather than simply selecting one of many video streams on your screen, instead you utilize the abstract view of the event (as viewed on your smartphone) to select your own personal viewpoint of the event. Therefore, you want answers to questions, such as: what are the positions of the Rabobank riders, what gaps and groups of riders are there on the track today, who is the virtual leader of the race at this moment, at what time can I expect the leader to pass my current position on the track, etc. Based upon

D. Roggen et al. (Eds.): EuroSSC 2008, LNCS 5279, pp. 206–219, 2008.

the answers to these questions you will move to the position which you decide will give you the best vantage point.

It is November 2007 in Gieten, Netherlands. It is cold and rainy, the perfect conditions for an international cyclo-cross race. The track lies partly in the woods and partly around a pond with steep banks. The track is about 3 km in length and each rider must complete 7 laps. The difference between the top riders and the ones that have a bad day is very big; after 2 laps the slowest rider is lapped by the fastest, but by then most distinguishing features of the riders are covered with mud – thus after 3 or 4 laps it is hard to see who is who. Using devices with the MIDAS middleware and race application preinstalled, attendees are able to see their own location and the location of the individual riders on a map of the track, the leading cyclist in the race, the total and remaining distance in meters, the gap between riders, as well as which riders are riding in the same group (so called gaps and groups analysis). All of this is shown live on your mobile device.

MIDAS (*Middleware Platform for Developing and Deploying Advanced Mobile Services*) [1] is a European research project concerning 3G and beyond, which aims to define and implement a platform to simplify and speed up the task of developing and deploying mobile applications and services. It is specifically designed to be used in MANETs. MIDAS enables applications running on different nodes to share information by inserting data in and retrieving data from a shared data space. This shared data space is implemented using a combination of data replication and remote operations – but this fact is transparent to applications. Therefore, for the purpose of this paper, we assume that all context information is available locally on a mobile device.

An application using this middleware calculates and displays gaps and groups of cyclists in near real time. This calculation needs to be performed as the cyclists' relative positions change, resulting in the synthesis of gaps and groups context. Moreover, the presentation on the display needs to be updated to reflect the current composition of groups. Thus, the middleware periodically obtains cyclists' geographic locations and utilizes information about the waypoints in the race. This data can be combined with the cyclists' data (such as name, team name, identification number, etc.), in order to perform context synthesis.

This newly generated context information can be in turn used by multiple applications. Hence, applications requesting customized context may share the cost of producing this synthesized context information. Additionally, each application need not be concerned with how this synthesis is implemented. However, some applications may want to implement their own synthesis functions. We refer to these functions as *context operators*. Context operators enable different applications and even different context-aware systems in the same domain to query each other about the context information which could be synthesized using the functions they implement. For example: a racing application and media application deployed on different devices should be able to remotely query each other (using the same middleware API and context operators) for results of the race and rankings of all athletes in the competition. The output of the operator is sent as a result of a context query. This result is called a *synthesized context*, since it was generated by context synthesis.

Our motivation and the idea for context synthesis using operators was previously presented in [2]. The main advantages of our approach are increased reusability, extensibility, and interoperability, facilitated by context operators and exploiting ontology based context modeling. This paper describes the realization of this approach via a race application developed on this middleware, and evaluates the context synthesis in terms of the response time to a context query sent by the application. The response time is divided into the time needed to find the correct operator, the time needed to obtain context information (formatted as ontology data) from its repository, and the time needed by this operator to perform the actual context synthesis.

The rest of the paper is organized in seven sections. Section 2 elaborates the MIDAS context modeling approach using ontologies for mobile devices. Section 3 presents our approach for context synthesis using context operators, while Section 4 describes the set of applications developed for sports scenario that illustrate the use of context operators for context synthesis. In Section 5 we give a performance evaluation. Section 6 provides a brief overview of related work. We conclude in Section 7 with a summary of the results and plans for future work.

2   Context Modeling Using Ontologies

In order for MIDAS to be a context aware framework it needs a mechanism for modeling and representing context. This context model must contain information specific to a specific domain of deployment of a MIDAS based service. A context model of a domain describes the people, objects, and relations between them which are typically encountered in a specific situation (or types of situations). Focusing only on a single domain makes it possible to create a very specific model, capable of representing a very fine level of detail, which otherwise could not be captured due to growing complexity of a more general model.

The context model used is an ontology, which is provided to the system in the form of a file. Thus the same middleware can be used in various domains given a new ontology file. We envision that an organizer of an event creates an ontology which represents the domain of this event. This ontology is provided to application developers (who can create applications for this particular domain). Once this ontology is created, other similar events can re-use the ontology, modifying it as required.

The ontology language chosen for the domain model is DL-Lite [3], which is a subset of OWL-DL optimized for fast reasoning on top of relational databases. This language supports the basic terms of classes and properties, and it handles statements about subsumption, disjointness, role-typing, participation constraints, nonparticipation constraints, and functionality restrictions. MIDAS implements an architecture for handling DL-Lite ontologies on a Java enabled mobile device [4].

The decision to use DL-Lite as a language for MIDAS ontologies was motivated by the results obtained during an initial experiment [5]. This experiment showed that using OWL-DL [6] with existing off the shelf solutions such as Jena [7] and Pellet [8] could **not** be applied on mobile devices, given their poor (slow) performance even on desktop machines and very high memory requirements. The use of existing ontology query languages, such as RQL [9] or SPARQL [10] was not analyzed. However, these solutions are usually not designed for mobile devices as their main focus is high expressivity. Thus, their practical usability in a mobile device setting is unlikely. We chose DL-Lite because of its relative simplicity and optimization for fast performance. The limitations in the description logic that made these improvements possible turn out not to be limiting when modeling a domain.

The syntax chosen for context model ontologies is the Manchester OWL Syntax [11] due to features which make it more suitable for applications on mobile devices than the usual OWL Syntax [12]. The main feature is that it is much easier to parse, as it requires only two linear scans of the ontology file, and does not require construction of a tree structure during parsing. Another feature is that an ontology is approximately half the size of the equivalent OWL representation, and because it is human readable it is possible to edit it by hand if necessary.

For representing the ontology on a MIDAS enabled mobile device we created a dedicated Lightweight Ontology library [13], which implements the Jena [7] API in a form suitable for mobile devices. This library parses the ontology file and creates an in-memory representation of the ontology (supporting all the structures present in OWL-DL) based on HashTables. Its simplicity suits resource constrained devices (such as J2ME mobile phones).

The scenario examined in this paper is a cycling race. Part of the context model ontology is shown in Fig. 1. This example shows only the part of the class hierarchy from the domain model, which contains classes corresponding to roles of users. Other classes (not shown) in the domain model are used to represent places encountered during the event, abstract entities such as a group of cyclists, a gap between two groups, etc.

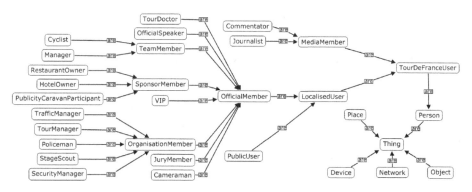

Fig. 1. Classes that describe roles of users involved in the cycling event

We consider five types of entities, which can be characterized as owners of context information: a person, a device, a network, a place, and an object. However, these entities are not independent, but have the following relations (see Fig. 2): a device is connected to a network; a person uses certain device(s); a person, device, and an object may be located at a place; a person and a device are somehow related to some other object(s). All entities are subclasses of the root class "Thing" in the ontology, from which all other terms are derived. Thus, we assign all context information to a certain entity and we can query information about an entity—i.e., user context, device context, network context, place context, and object context.

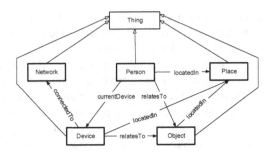

Fig. 2. Context entities and their relations in the context model

The context modeling architecture is implemented by a Context Knowledge Base component in the middleware. The API offered by this component makes it possible to model context by means of objects of the type *DomainInstance*, each of which represent physical entities. A *DomainInstance* can be added or removed as needed. Each *DomainInstance* object can have a number of property values assigned to it, and can belong to a number of classes. These classes are represented by objects of the type *DomainClass* and *DomainProperty* (respectively) which correspond to those present in the domain model ontology.

Context information needs to be stored by the middleware before it can be queried or synthesized. In order to store, retrieve, and manipulate the formatted (higher-level) context information, we developed a means of mechanically mapping the domain classes from the context model to the corresponding java classes, as well as from property names to java class variables.

3   Context Synthesis Using Context Operators

Operators for context synthesis are domain-specific functions over the context data. The benefits of these operators are that by performing operations over the existing context information, new context information that previously did not exist in the system can be produced. The output of the operation performed by an operator, a *synthesized context*, is sent to the application as a result of a context query. Operators could be used on a higher level to synthesize information of a certain user, device, network, place, or other object, as illustrated earlier in Fig. 2.

Operators are bundles of both a description and implementation; and described by an ontology, similar to the representation of context. They are implemented as java scripts that perform an action specified in the operator's ontology. The operator's description specifies the name of this operator, the types of the required input arguments, the returned output type, and the list of other operators used in performing the operator's function. As with the context model, operators are created for a specific domain and can be used by a set of applications in that domain. In order to provide context synthesis functions for applications in another domain, a new set of operators needs to be provided to the middleware, along with their ontology schema.

We distinguish between generic and specialized operators. Generic operators are part of an ontology schema, representing an umbrella for all the different implementations of a function they provide. They are also part of an API provided to application developers. On the other hand, specialized operators can be created/modified and inserted into the middleware by application or system developers. Specialized operators are not directly visible to application developers; which operator is invoked will be determined by the middleware at runtime.

Specialized operators are implemented as scripts using Beanshell [23], an open source java script engine. In our implementation the operator scripts are part of the context service process and they can be programmatically added and removed by the middleware.

Fig. 5 shows the structure of the *Operator space* – a repository of operators. The root folder (i.e. operators/) contains all generic operators (which are also folders), containing in turn their specialized operators. Note that specialized operators are bundles of an operator description (an instance of the operator ontology encoded in Manchester OWL format, i.e., a .man file) and an operator implementation (a java script written in Beanshell, i.e., a .bsh file).

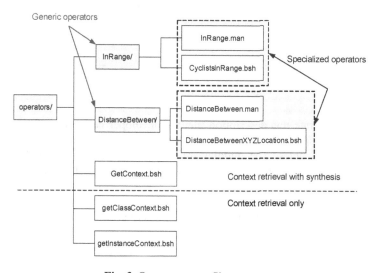

Fig. 3. Operator space file structure

The root folder of the *Operator space* shown in Fig. 3 also contains three specific operator scripts which are responsible for retrieving context data of the specified context owner, from the Context knowledge base: *GetContext.bsh, GetClassContext.bsh*, and *GetInstanceContext.bsh*. Note that they do not have a generic operator representing them, and they are used for distinct purposes. As previously noted, when specific context operators need to retrieve context, they will provide *DomainInstance* objects to the *GetContext* operator to retrieve the missing context values. It is also possible to retrieve context data directly from the repository without context synthesis, via the *GetClassContext* and *GetInstanceContext* scripts. *GetInstanceContext* is used to obtain the domain instance with the supplied datatype properties from the context query.

We can also query the Context knowledge base for other properties of the same instance. *GetClassContext* is used when we do not know the instance, but rather use a domain class with the specified property name-value pair to identify this instance.

An example of an operator description file, *InRange.man* is presented in Fig. 4. This file contains all the specialized operator descriptions. Fig. 3 shows only *CyclistsInRange*, but there could be others as well (e.g., *UsersInRange*). The description of the *CyclistsInRange* (specialized) operator is interpreted in the following way: it has the name "*CylistsInRange*" and is derived from a generic operator (i.e. *InRange*). It requires an input of the type Cyclist and produces an output value of the type Cyclist. The operator uses the result from another (simpler) operator *DistanceBetweenXYZLocations* to calculate the distance between two locations.

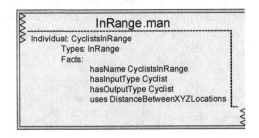

Fig. 4. CyclistsInRange description

In order to generate this description file, the developer needs to programmatically set the type of this specialized operator, the list of input types, the output type, as well as operator dependencies. The middleware will automatically append this operator description to the correct ontology file (if this file does not exist, it will be created in the correct location).

Note that all specialized operator scripts take as inputs *DomainInstance* objects, which are instances of classes specified as input types in their operator description file. Thus these domain instances pass the input arguments from the context query to the operator's method, and can be used to retrieve the missing information from the Context knowledge base (if needed).

3.1 Operator Matching

The context synthesizing process determines the most appropriate specialized operator to invoke from the available (specialized) operators by using a reasoning process (which takes into account the required output type and supplied input types). The idea behind the operator matching algorithm, illustrated in Fig. 5, is to enable different applications (or even different context systems) in the same domain (in our scenario a sport domain) to use the same "functions" to synthesize context information, without being concerned about the implementation of these functions. The operator matching algorithm returns the specialized operator with either exactly the same description as specified by the query or a more generic one.

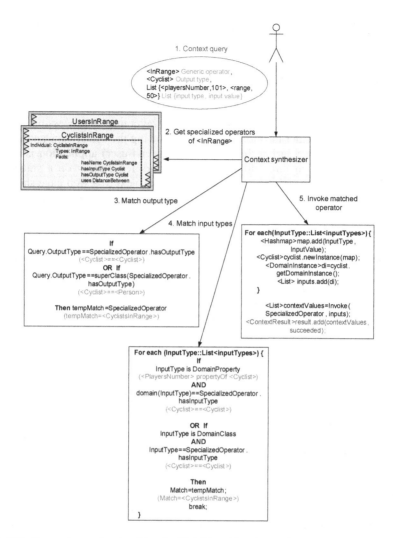

Fig. 5. This figure shows the algorithm itself, initiated by the user's context query, along with the invocation of the matched (specialized) operator

An example of a context query is: *InRange("101", 50, ModelConstants.Cyclist),* where the response time is bounded to *5sec.* This example can be interpreted as follows: retrieve all cyclists in the range of 50 meters from the cyclist with the ID="101" and the result should be returned within 5 seconds. If the result is not computed by that time, the synthesis process will be interrupted, and a response will be returned to the query initiator containing an empty list of values and a flag indicating that the query was unsuccessful. After receiving the query, the operator matching algorithm retrieves all available specialized operators and processes the supplied data in order to find an exactly matching specialized operator (by checking if output and input types of the operator and the query match). Otherwise it will return a more generalized one, i.e. *UsersInRange*, which would return *Users* instead of *Cyclists* as result. Finally, it invokes the matching operator.

4 Cyclist Race Application

The cyclist race application set consists of a number of applications responsible for: 1) entering static cyclist data and managing track waypoints, 2) processing and showing a list of the latest rider location data and 3) showing the actual gaps and groups of cyclists to the end user during the race. The last (end-user) application is available with a user interface in three different form factors for display on a small device (Nokia N800), a laptop (HP tablet), and as a side bar next to a web page shown on a laptop. The processing application was at the time of the race not actually deployed on a mobile device; however, in Section 5 we give an evaluation of context synthesis on a Nokia N800.

The geographic locations of the cyclists are obtained from GPS receivers attached to cyclists' arms and this context is synthesized into *gaps and groups* information. The *gaps and groups* information is in turn broadcasted to all interested users equipped with the MIDAS middleware and an end-user application installed on their devices. The frequency of updates is about once every three seconds. A video demonstrating the live race at the Super Prestige Cyclocross using MIDAS middleware and the described application set can be seen at [21].

The gaps between groups of cyclists and the composition of groups are synthesized from the following cyclists' context: the last known cyclists position information, the roadbook waypoints, and the configured maximum distance between two consecutive cyclists of one group. A gap is defined as a distance between locations of two consecutive cyclists that exceeds a predefined threshold. In cycling a distance of about 25 meters is considered a gap. Cyclists between two consecutive gaps compose a group. In order to calculate gaps and groups, the application needs to calculate the distance between the successive waypoints and their distance to the finish (based on Haversine's Formula [14] combined with John P. Snyder's curvature [15]).

Figure 6 illustrates the operators used to calculate gaps and groups information. In order to improve the performance, all real time objects are stored in the memory. To share these objects between applications singleton instances are stored in the operator space running environment; therefore an operator has to be used to interact with these objects. Moreover, the output of one operator is used as an input to the other one.

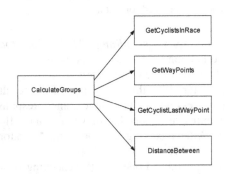

Fig. 6. Set of operators used in the application

To calculate the gaps and groups, the algorithm exploits the fact that every cyclist cycles from one waypoint to the other and sends several GPS measurements while on this path to reach the waypoint. Based on received GPS measurements, the algorithm computes location, distances between cyclists, their order, and if the distance between two cyclists is 25 meters or greater, then there is a gap.

Groups in the race are presented graphically to the user via a user application, as shown in Fig.7. The circle represents the whole track. Each dot represents a group. The progress of the groups is shown relative to Start and Finish. Additionally, the list of groups is presented to the user as a textual table. The first column of the table contains the group name (1 to n), the second column shows the number of cyclists present in the group, and the third one contains the distance to the preceding group. The leading group distance is replaced with a "Leading" indicator. For every cyclist, the first and last name, player number, as well as distance to the finish are shown. Once a group finishes the race, the distance is replaced by a "Finished" indicator, the cyclist icon is replaced by a flag, and the line is printed in green.

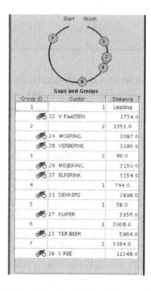

Fig. 7. Application GUI with actual data

5 Performance Evaluation

The MIDAS middleware and applications are implemented in Java. We ran all performance tests on a Nokia N800 device with the JamVM virtual machine [22] with a compiler for Java 1.4. This device was chosen by the MIDAS project because it is Linux based, allowing network and low-level programming. We also used a third party library for implementation of java scripts, Beanshell [23].

The performance of context synthesis is evaluated in terms of the response time of operator matching, context retrieval, and context processing (i.e., operator invocation),

Table 1. Response times

Average response times with varying number of specialized operators (i.e., 1, 2, 5, 10)	Based on 10 **first** queries	Standard deviation (based on 10 **first** queries)	Based on 10 **subsequent** queries	Standard deviation (based on 10 **subsequent** queries)
Matching algorithm time	2.49 sec	0.009 sec	1.94 sec	0.07 sec
Loading specialized & root scripts time	1.7 sec	0.087 sec	No average, for the first time only (1.7 sec)	No standard deviation
Total operator matching time	4.2 sec	0.087 sec	1.94 sec	0.07 sec
Context retrieval time	0.37 sec	0.006 sec	0.09 sec	0.001 sec
Loading dependency scripts time	0.15 sec	0.001 sec	0.17 sec	0.015 sec
Operator invocation time	0.67 sec	0.008 sec	0.36 sec	0.04 sec
Total query time	5.4 sec	0.045 sec	2.57 sec	0.07 sec

as well as the overall response time to a query sent by an application. The values shown in Table 1 were obtained by sending the same context query, but varying the number of available specific operators (i.e., 1, 2, 5, and 10) when performing the operator matching algorithm, and then calculating the mean value.

Note that before the java scripts can be invoked, they have to be loaded into the interpreter and the classpath has to point to the folder where these scripts reside. These scripts can also invoke other scripts (from different folders), thus these other scripts need to be invoked in the caller's context (the so called *namespace*). Therefore, when the first query is sent, the total time needed to find the most appropriate specialized operator (i.e., the *total operator matching time*) also includes the time needed to set the namespace to point to the generic operator folder (e.g., *InRange*), as well as load specific operator scripts from this folder and from the root operator folder. For all successive queries this operation is cached. When invoking the specialized operator found by the matching algorithm, some additional time is needed to load the scripts from the dependency operator folder (e.g., *DistanceBetween*).

As it can be seen from Table 1, the response times for the first query are twice as large as for the other following queries, because the caching speeds up the subsequent operations. The operator matching algorithm takes 2 seconds on average, however for the first query it requires 4 seconds (including the initial time needed for loading the necessary scripts). Context retrieval (of three cyclists' data) was rather quick as was the operator invocation time. The number of concepts required by an application was small. With regards to performance with increasing number of domain instances, please refer to [4]. Note that operator invocation time includes the time needed to invoke *CyclistsInRange* and *DistanceBetween* operators. We used SQL prepared statements to retrieve context from an HSQL database. The total time needed to receive the result of context query took on average 2.5 seconds, but 5.4 seconds for the first query.

Note also that in some other scenario it could happen that after the second query the first query is made again but containing some other operator, this will also require operator matching. However, we plan (as future work) to introduce caching of queries and matched specialized operators in order to reduce the total query time.

There were 1000 spectators along the race course. Note that this deployment was intended as a proof of concept to validate middleware functionalities and was not designed to be an evaluation of the system using a statistically significant number of users. However, the impression of 9 users (monitoring the race on 6 tablet PCs and 3 Nokia N800 devices) was very positive. A few seconds of delay did not affect their "near real-time experience". Furthermore, users liked the way that they could select their favorite cyclists in the application, in order to know when he/she will pass their location. Zooming functionality also helped to improve overcome the limitations of the small screen when more cyclists were tracked during the race.

So far we have not examined the cases when context changes rapidly nor we have considered the issues concerning uncertainty in the context. We plan to address these issues in future work.

6 Related Work

Our context synthesizing work was inspired by the Aura Contextual Information Service (CIS) research project [16]. However, our queries are not SQL-like, but instead they are object-oriented, containing context operators which perform synthesizing operations. Context operators can in turn use other simpler operators to execute smaller tasks and to reuse existing functionality.

Modeling context with ontologies is in itself not a novel idea. Surveys of context modeling frameworks clearly indicate that modeling context with ontologies is the most expressive way to do it [17]. Typically, mobile devices being part of a context aware system need to remotely access the ontology model, and the context data. In case of CoBrA [18] the remote facility is an 'intelligent agent', called a Context Broker, which acts as a central point of the system maintaining a representation of context common to all the devices in the network. The SOCAM [19] solution also relies on a shared context space located on an external device (an OSGi gateway) which can be accessed by multiple context aware services. MIDAS differs from these architectures in that it is capable of handling ontologies on mobile devices, which makes it possible to provide local access to context modeled with ontologies for every device in the network. This seems especially useful in ad-hoc networks where access to a central server cannot be provided.

J. I. Hong and J. A. Landay [20] emphasize a need for creating a basic infrastructure services and application-specific services, the latter implemented on a case-by-case basis. One such basic infrastructure service is automatic path creation, which transforms raw sensor data to higher-level context data. It automatically composes operators based on high-level needs and what resources are available. Our work extends this idea to enable multiple applications or even different context-aware systems to use the **same** operators designed for a specific domain without being concerned about their implementation. Moreover, we also enable chaining of operators, where each operator takes some existing context information (as defined by a context model) as input and provides new context information as an output. All applications can reuse already deployed operators and add their own implementations of the same generic operators.

7 Conclusion and Future Work

We have presented and evaluated the approach for context synthesis using operators on a Nokia N800 device. Operators for context synthesizing are domain-specific functions over the context data. The benefits of these operators are that by performing operations over the existing context information, new context information that previously did not exist in the system can be produced. Moreover, applications can use the **same** operators to synthesize context information, without being concerned about their implementation. This also enables applications to share the cost of context synthesis by querying about the result of operators invocation.

We have evaluated this operator-based context synthesis approach in terms of response time to context query sent by the application and showed that it is possible to perform context synthesis operation in near real time (i.e., with the average latency of 2 seconds) on the mobile device. The main advantages of context operators are the reusability, extensibility, and interoperability, facilitated by ontology-based context modeling. For this purpose MIDAS provides a dedicated Lightweight Ontology library for representing and manipulating ontologies on mobile devices. We also demonstrated the use of context operators in the cyclist race application.

We plan to evaluate the response time of executing the remote operator invocation as well as to use caching decisions made by operator matching algorithm for a certain context query. We will also conduct a user study based on our next deployment. As a next step in context synthesizing we plan to use operators to combine inference algorithms in order to derive about high-level context.

Acknowledgements. The authors of this paper would like to acknowledge the partial financial support given to this research by the EU IST MIDAS project (6th Framework Programme, contract number 027055). We would also like to thank Prof. Gerald Q. Maguire Jr. for his valuable comments to this research work.

References

1. EU FP6 IST MIDAS project (2008), http://www.ist-midas.org
2. Devlic, A., Klintskog, E.: Context retrieval and distribution in a mobile distributed environment. In: Third Workshop on Context Awareness for Proactive Systems (CAPS 2007), Guildford, UK (2007)
3. Calvanese, D., De Giacomo, G., Lembo, D., Lenzerini, M., Rosati, R.: DL-Lite: Tractable description logics for ontologies. In: 20th National Conference on Artificial Intelligence (AAAI 2005), Pittsburgh, Pennsylvania, USA, pp. 602–607 (2005)
4. Koziuk, M., Domaszewicz, J., Schoeneich, R.O.: Mobile Context-Addressable Messaging with DL-Lite Domain Model. In: Roggen, D., Lombriser, C., Tröster, G., Kortuem, G., Havinga, P. (eds.) EuroSSC 2008. LNCS, vol. 5279, pp. 168–181. Springer, Heidelberg (2008)

5. Domaszewicz, J., Koziuk, M., Schoeneich, R.O.: Context-Addressable Messaging with ontology-driven addresses. In: The 7th International Conference on Ontologies, Data-Bases, and Applications of Semantics (ODBASE 2008), Monterrey, Mexico (to appear, 2008)
6. Smith, M.K., Welty, C., McGuinness, D.L.: OWL Web Ontology Language Guide. W3C Recommendation (2004), http://www.w3.org/TR/owl-guide/
7. Carroll, J.J., Dickinson, I., Dollin, C., Reynolds, D., Seaborne, A., Wilkinson, K.: Jena: Implementing the semantic web recommendations. Technical Report HPL-2003 (2003), http://citeseer.ist.psu.edu/carroll04jena.html
8. Sirin, E., Parsia, B., Grau, B.C., Kalyanpur, A., Katz, Y.: Pellet: A practical OWL-DL reasoner. Web Semantics: Science, Services and Agents on the World Wide Web 5(2), 51–53 (2007)
9. Magkanaraki, A., Alexaki, S., Christophides, V., Plexousakis, D., Scholl, M., Tolle, K.: RQL: A Functional Query Language for RDF. In: Gray, P.M.D., Kerschberg, L., King, P.J.H., Poulovassilis, A. (eds.) The Functional Approach to Data Management: Modelling, Analyzing and Integrating Heterogeneous Data. Springer, Heidelberg (2004)
10. Sirin, E., Parsia, B.: SPARQL-DL: SPARQL Query for OWL-DL. In: Proceedings of the OWLED 2007 Workshop on OWL: Experiences and Directions, Innsbruck, Austria, June 6-7 (2007)
11. Horridge, M., Drummond, N., Goodwin, J., Rector, A., Stevens, R., Wang, H.: The Manchester OWL Syntax. In: OWL: Experiences and Directions 2006, Athens, Georgia, USA (2006)
12. Patel-Schneider, P.F., Hayes, P., Horrocks, I.: OWL Web Ontology Language Semantics and Abstract Syntax. W3C Recommendation (2004), http://www.w3.org/TR/owl-semantics/
13. Jabłonowski, M., Boetzel, P.: Middleware Layer For Semantic Object Tagging. Master Thesis at Warsaw University of Technology, Warsaw, Poland (2007)
14. Sinnott, R.W.: Sky and Telescope. Virtues of the Haversine 68(2), 159 (1984)
15. Snyder, J.P.: Map Projections – A Working Manual., U.S. Geological Survey, Professional Paper 1395, US Government Printing Office, Washington DC (1987)
16. Judd, G., Steenkiste, P.: Providing Contextual Information to Pervasive Computing Applications. In: First IEEE International Conference on Pervasive Computing and Communications (PerCom 2003), Fort Worth, Texas, pp. 133–142 (2003)
17. Strang, T., Popien, C.L.: A context modeling survey. In: Workshop on Advanced Context Modeling, Reasoning and Management as part of the 6th International Conference on Ubiquitous Computing (UbiComp 2004), Nottingham, England, pp. 33–40 (2004)
18. Chen, H., et al.: A Context Broker for Building Smart Meeting Rooms. In: Proceedings of the Knowledge Representation and Ontology for Autonomous Systems Symposium, 2004 AAAI Spring Symposium, Palo Alto, CA, USA (2004)
19. Gu, T., Wang, X.H., Pung, H.K., Zhang, D.Q.: An Ontology-based Context Model in Intelligent Environments. In: Proceedings of Communication Networks and Distributed Systems Modeling and Simulation Conference, San Diego, California, USA (2004)
20. Hong, J.I., Landay, J.A.: An Infrastructure Approach to Context-Aware Computing. Human-Computer Interaction 16(2, 3, 4), 287–303 (2001)
21. MIDAS video (2007), http://www.youtube.com/watch?v=yulUmlVH8Jc
22. JamVM – A Compact Java Virtual Machine (2008), http://jamvm.sourceforge.net/
23. Beanshell - Lightweight scripting for Java (2008), http://www.beanshell.org/

Using Aesthetic and Empathetic Expressions to Motivate Desirable Lifestyle

Tatsuo Nakajima, Hiroaki Kimura, Tetsuo Yamabe, Vili Lehdonvirta*,
Chihiro Takayama, Miyuki Shiraishi, and Yasuyuki Washio

Department of Computer Science and Engineering
Waseda University
*Helsinki Institute for Information Technology
tatsuo@dcl.info.waseda.ac.jp

Abstract. In recent years, the deteriorations of living habits like immobilization or unhealthy diet are becoming serious social problems in many developed countries. Even if we know the importance, it is difficult to change our undesirable habits and to maintain a desirable lifestyle. This study demonstrates a concept called *ambient lifestyle feedback systems* to be used to motivate people to change their undesirable habits to improve their lifestyle. In the concept, aesthetic and empathetic expressions reflect the feedback of the user's current behavior to the user. When keeping desirable habits, the user is offered with a feedback expression designed to boost his positive emotion. When turning to undesirable habits, the feedback expression is changed to increase the user's negative emotions. In this paper, we present brief overviews of four case studies of ambient lifestyle feedback systems, and discuss several findings that we while designing and evaluating the case studies. Future directions will also be discussed.

1 Introduction

Although we know that it is important to keep desirable habits with a great effort, we are apt to being lazy, and having an easy life. Several previous approaches [11,12] have tried to change a user's daily habits to motivate a better lifestyle. Also, persuasive technologies propose several principles how computing technologies are used to change human behavior and attitude [5]. We have designed our own solution to motivate the desired changes of our behavior. The solution, called *ambient lifestyle feedback systems* [15], is presented as a set of three design principles [15]. We used some basic tenets from operant conditioning [4] as a basic principle for changing a user's habits. The most obvious issue is that the system should include a feedback loop between the user's behavior and the expression shown on an ambient display.

Expression has a role not only as a representer of information but also as an external motivator for the user's future action. A picture is originally designed to be watched and to have something attractive, but it is also suitable for information visualization [8]. One of the important challenges in ubiquitous computing

D. Roggen et al. (Eds.): EuroSSC 2008, LNCS 5279, pp. 220–234, 2008.

is how to represent large amounts of information in a calm and unobtrusive manner. Aesthetic expression of the information is also important in order to accept ubiquitous computing technologies in our daily lives. Our approach enhances traditional ambient displays to claim that an ambient display can be used to persuade to change a user's undesirable habits, and keep desirable habits.

In this paper, we discuss how aesthetic and empathetic expressions are effective in persuading a user to change his undesirable habits. The emotional impact is very important in making a user to keep desirable habits although he considers it to be hard and challenging. For example, "quit meters" [22] provide smokers with constant feedback on how much money is wasted and how many minutes of life are lost. "Carb counters" [21] for Palm handheld devices provide instant feedback on meal choices. Compared to games, quit meters and carb counters do not require setting aside time for game sessions, as their use is intended to happen during normal daily activities. But the feedback they provide lacks the engagement and fun that games provide, lessening their emotional impact.

In Section 2, we describe an overview of ambient lifestyle feedback systems. Section 3 presents four case studies, Persuasive Art, Virtual Acquarium, Mona Lisa Bookshelf, and EcoIsland. In Section 4, several findings are described, and Section 5 concludes the paper, and shows future directions.

2   Ambient Lifestyle Feedback Systems

Ambient lifestyle feedback systems as shown in Fig. 1 are intended to be implemented using ubiquitous computing techniques, including sensors and ambient display devices, but most implementation details are determined by the needs of a particular application and what behavior it targets to satisfy the following principles below. To lend a solid framework for feedback design, we referred to elementary behavioral psychology. Behavioral psychology is a discipline dealing with the relationship between behavior and consequences. It posits that the form and frequency of behavior can be affected by controlling the consequences. The principle to design the systems consists of the following three components.

Passive observation: One of the key factors limiting the applicability of the earlier solutions to our intended purpose is the various burdens they place on a user, either in the form of time use or effort. To avoid the burdens of self-reporting, the system should be able to passively observe the user's behavior. To eliminate the need to set aside time or go to a special place, the system should be integrated with normal daily activities. Thus our first design principle is to use observations of the users' behavior as the system's input, as opposed to using keystrokes or some other proxy behavior. This also facilitates the delivery immediate feedback, a key factor in the effectiveness of operant conditioning.

Emotional Engagement: The fact that the feedback is delivered in a non-disruptive way must not mean that it ends up being irrelevant to the user. To effect a change in behavior in operant conditioning, we must be able to administer some sort of meaningful consequences to the user. We do not have means to effect

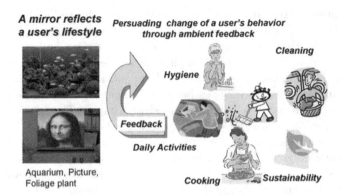

Fig. 1. The ambient lifestyle feedback system could be described as a kind of mirror, because it reflects something about the user. The mirror does not show the usual outward appearance, but reflects more personal facts that may otherwise go unnoticed.

changes in physical reality (such as threatening the user's family), so would it be possible to make the user care about changes in the internal state of a computer system? Computer games seem to be able to do this. Good games are able to provoke a range of emotional responses, from fun and satisfaction to guilt and discontent. By mimicking the techniques used in computer games, we should be able to build an emotionally engaging feedback system, allowing us to administer punishments and rewards without any physical resources. By "emotional engagement" we do not necessarily mean strong and deep emotional responses, but the simple kicks that make many games interesting and addictive.

Ambient Feedback: To complete the integration of the system into the user's daily living environment, we must also make sure that the output produced by the system is appropriate. We refer to Mark Weiser's concept of calm technology: technology that is able to leverage our peripheral perception to deliver information, as opposed to constantly demanding direct attention [19]. Ambient lifestyle feedback systems should be designed to blend into their environment and to be able to deliver information in the periphery. A loud or disruptive feedback system might even find itself thrown out of the house or workplace.

3   Four Case Studies

3.1   Persuasive Art

Decorating walls with pictures is common at home. Pictures are a very important way to increase aesthetic feeling in our daily lives. Persuasive Art uses a painting to motivate a user to walk at least 8000 steps every day to keep his/her fit. The number of steps are monitored automatically and stored into a computer. The painting shows the feedback of the current status of the user's exercise to motivate him to maintain desirable habits. Persuasive Art currently offers the following four paintings as shown in Fig. 2. The landscape painting includes a tree that grows and withers, The figure painting is Mona Lisa, the abstract painting has objects that change in size, and the still life picture contains a changing number of orbs.

When using the landscape painting, the tree's growth is varied according to the user's behavior. When the user maintains desirable habits, the tree will grow, but if he stops the desirable habits, the tree will get sick. The painting adopts the following metaphor. The increase of healthy activities makes the tree more healthy, but the neglect of the exercise makes the tree sick.

Fig. 2. Four alternative virtual paintings

Motivating humans can be classified into two approaches. One is to make users aware of their current situation and the other is to enhance the user's willingness to change his habits. Motivating a change of habits can also be classified into two types. The positive expression style increases a user's positive emotion to motivate a change in the user's undesirable habits. The user feels happy when changing his/her undesirable habits even if the change is challenging and hard. Another type is the negative expression style. This promotes negative emotion to feel a sense of crisis that motivates to change the user's undesirable habits. For instance, if a user looks at himself in a mirror and finds that he is significantly overweight, this may motivate him to do more exercise.

3.2   Virtual Acquarium

Virtual Aquarium shown in Fig. 3 has the objective of improving users' dental hygiene by promoting correct toothbrushing practices. The system is set up in the lavatory where it turns a mirror into a simulated aquarium. Fish living in the aquarium are affected by the users' toothbrushing activity. If users brush their teeth properly, the fish prosper and procreate. If not, they are weakened and may even perish.

In this system, we used a 3-axis accelerometer that is attached to each toothbrush in a household. Since toothbrushes are usually not shared and each Cookie

Virtual Aquarium
When he starts toothbrushing, fishes in the aquarium start happily dancing.

Cookie
The toothbrush embeds *Cookie* that is a coin-size wireless customizable sensor device.

Fig. 3. A user toothbrushing in front of a Virtual Aquarium. The toothbrush carries a wireless sensor device that tracks the users' activity.

has a unique identification number, we are able to infer which user is using the system at a given time. Toothbrushing patterns are recognized by analyzing the acceleration data. Fig. 3 shows a user brushing his teeth in front of the Virtual Aquarium using a brush with a Cookie attached. The toothbrush is able to observe how a user brushes his/her teeth passively without requesting extra actions to play the game.

The objective of Virtual Aquarium is to promote good toothbrushing practices. In this prototype, the ideal behavior was defined as follows: 1) users should brush their teeth at least twice per day; 2) one session should involve at least three minutes of brushing; and 3) brushing should involve patterns that ensure the teeth are properly cleaned. User behavior is compared to this ideal and translated to feedback as described below. We believe that the existence of an aquarium is in-artificial in a lavatory, but the aquarium enriches our daily life.

Fig. 4. Screen Images of Virtual Aquarium

When a user begins to brush her teeth, a scrub inside the aquarium starts cleaning algae off the aquarium wall. At the same time, a set of fish associated with the user starts moving in the aquarium in a playful manner. When the user has brushed for a sufficient time, the scrub finishes cleaning and the fishes' dance turns to a more elegant pattern. When the user finishes brushing, the fish end their dance and resume their normal activities. Both the activity of the fish and the movement of the scrub are designed in such a way as to give the user hints regarding the correct method of toothbrushing. The left picture in Fig. 4 shows a scene from the aquarium during brushing. However, if a user does not brush his/her teeth sufficiently, the aquarium becomes dirty, and the fishes in the aquarium become sick.

The fish's health is visibly affected by how clean the aquarium is. If a user neglects to brush her teeth, some fish fall ill and may even die. In contrast, faithful brushing may result in the fish laying eggs (The right picture in Fig. 4). At first, the eggs are not very likely to hatch. If the user continues to brush consistently for a number of days in row, the incubation ratio increases. This way, the accumulated feedback gives clues to the correct behavior and attempts to maintain motivation over a period of time.

3.3   Mona Lisa Bookshelf

Resources shared by a number of people, such as a public toilet or a bookshelf in a research lab, tend to deteriorate quickly in a process called the tragedy of the commons. This happens because each individual derives a personal benefit from using the resource, while any costs are shared between all the users, leading to reckless use. Garret Hardin, the ecologist who popularized the concept, noted that this belongs to the category of problems that cannot be solved by technology alone, requiring instead a change in human behavior [9]. Mona Lisa Bookshelf, is aimed at keeping a bookshelf organized. It tries to encourage users to keep books in order and to return missing books, but also to take books out every now and then for reading. Each book in the shelf is linked with a piece of a digital image of the Mona Lisa. Like a picture puzzle, the image changes according to how the books are positioned. A high-quality flat display placed near the bookshelf shows the image to the users.

Fig. 5. Mona Lisa Bookshelf prototype installation

The tracking of a user's behavior is based on optically detecting books in the shelf. In the prototype system, visual tags are attached to the spines of the books to facilitate their detection and identification. Visual tags are also attached to the corners of the shelf to determine its dimensions (Fig. 5(Left)). The detection system (Fig. 5(Middle)) comprises the following hardware: a digital video camera (iSight by Apple), a high-resolution digital camera (D50 by Nikon) and two infrared distance detectors (GP2D12 by SHARP). The distance sensors and the digital video camera are used to detect whether a user is manipulating books in the shelf. OpenCV, a real-time computer vision software library, is used to analyze the video signal. As soon as a user is seen leaving the shelf, the high-resolution still camera takes a picture of it and all the books contained within it. Images captured by the still camera are analyzed by the VisualCodes [17] software library, which recognizes the visual tags attached to the books. The system is shown installed in Fig. 5(Right). Each visual code yields data regarding its position, alignment and identity. This is then translated into context information that describes the bookshelf's width and height, which books are currently contained in shelf, and how they are aligned and ordered. This information is then passed to the feedback logic component. The above approach is able to observe how a user uses her bookshelf passively without requesting extra actions to play the game.

In this system, the feedback logic aims to encourage the following ideal behavior: 1) books should be arranged correctly and aligned neatly; and 2) at least one of the books should be read at least once per week. The correct arrangement of the books is pre-programmed, and could be e.g. alphabetical. User behavior is compared to this ideal, and translated to feedback as described below.

Mona Lisa Bookshelf also offers two expression styles to return feedback to a user to encourage cleaning his/her bookshelf or reading books in the following ways. When a book is removed from the shelf, the corresponding piece of the Mona Lisa image also disappears. If books are lying on their face or otherwise misaligned, the pieces of the image also become misaligned, distorting the picture. When the books are arranged neatly, Mona Lisa smiles contently. The assumption is that users are aware of how da Vinci's Mona Lisa is supposed to look like, and as when completing a picture puzzle, inherently prefer the correct solution to a distorted image. The feedback thus provides clues and motivation for keeping the bookshelf organized. The left picture in Fig. 6 shows and example of a distorted image. Also, if none of the books are removed from the shelf for over a week, Mona Lisa starts getting visibly older. The right picture shows and example of an aged portrait. As soon as one of the books is removed from the shelf, she regains her youth.

3.4   EcoIsland

Global warming caused by greenhouse gases released into the atmosphere through the actions of man is believed to be a major threat to the earth's ecology [10]. Efforts to reduce greenhouse gas emissions come in two forms: technological solutions and changes in human behavior. Technological solutions broadly include improving energy efficiency and developing cleaner energy sources. Dramatic changes in human behavior may also be necessary if catastrophic climate change is to be avoided.

Public and private efforts to change individual behavior towards more environmentally friendly practices usually rely on education, but there are psychological limits to the ability of education alone to effect behavioral change. Even when a

Fig. 6. Two example outputs of the Mona Lisa Bookshelf. The image on the left shows that some books are tilted and in the wrong order. Some books are also missing. The image on the right side indicates that none of books have been picked up for a long time.

person full-well knows that a particular behavior is detrimental enough to their long-term well-being to offset any possible short-term benefits, their may still irrationally choose the short-term indulgence. Future consequences, while widely known, are easily ignored in the present.

EcoIsland is a game-like application intended to be used as a background activity by an ecologically minded family in the course of their normal daily activities. A display installed in the kitchen or another prominent place in the household presents a virtual island. Each family member is represented on the island by an avatar (Fig. 7). The family sets a target CO_2 emission level (e.g. national average minus 20%) and the system tracks their approximate current emissions using sensors and self-reported data. If the emissions exceed the target level, the water around the island begins to rise, eventually sweeping away the avatars' possessions and resulting in a game over.

On their mobile phones, the participants have a list of actions that they may take to reduce the emissions: turning down the air conditioning by one degree, taking the train instead of the car, et cetera. Upon completing an action, a participant reports using the phone, and the water level reacts accordingly. Reported activities are also shown in speech bubbles above the corresponding avatars. A lack of activity causes the avatars to suggest actions. Participants can also see neighboring islands and their activities in the display, and can list buy and sell offers for emission rights on a marketplace. Trading is conducted using a virtual currency obtained from a regular allowance. The credits are also used to buy improvements and decorations to the island, so successful sellers can afford to decorate their island more, while heavy emitters have to spend their allowance on emission rights.

The general approach from ambient lifestyle feedback systems is to provide a feedback loop for user behavior. The virtual island shown in the display acts as a metaphor and makes the participants conscious of the ecological consequences of their choices and activities. We also tap into social psychology, attempting to exploit *social facilitation* and *conforming behavior* to encourage the desired

Fig. 7. Some Screenshots of EcoIsland

behavior. Social facilitation is the phenomenon where a person performs better at a task when someone else, e.g. a colleague or a supervisor, is watching [20]. Conforming behavior is the desire not to act against group consensus [1]. EcoIsland's design facilitates these by involving the whole family, and by presenting the participants' activity reports in the speech bubbles and providing contribution charts and activity histories. On the other hand, the fact that the game is played by a family unit instead of an individual means that participants can also agree to assign tasks to certain members.

Lastly, there is the trading system, which is based on the same principle as industry level emissions trading systems: reductions should be carried out in places where it is easiest to do so. A family that finds it easy to make significant reductions can sell emission rights to households that find it difficult due to e.g. location or job. This should make it possible to attain the same amount of total reductions with a lower total cost (measured in disutility), promoting use of the system.

4   Some Experiences with Case Studies

4.1   Sensing and Lightweight Interaction

Virtual Aquarium uses a 3D accelerometer to recognize the movement of the user's toothbrush to observe the users behavior without his explicit interaction. Our experiences show that recognizing the users behavior with sensors has its limitations in reliability. In Virtual Aquarium and Mona Lisa Bookshelf, we chose to analyze a very simple context that can be implemented in a reliable way. It is very difficult to analyze the user's behavior correctly without heavy-weighted algorithms. Thus, EcoIsland uses a self-reporting method to input what kind of actions the user takes in order to avoid complex behavior analysis.

EcoIsland encourages the user to input his actions to reduce CO_2 emission since he is recognized as an eco-conscious person. However, we believe that lightweight interaction techniques such as using gestures is important in order to compensate implicit interaction for realizing passive observation. The user will be able to input his current action with a minimum cognitive effort. Lightweight interaction can also be used to correct mistakes of behavior analysis. Gesture analysis is easier to compare than general behavior analysis. The combination of explicit interaction with gesture analysis and implicit interaction with sensor data analysis is a very interesting topic and has a critical role in our future researches.

One of the problems in the current case studies is that a user may cheat the analysis of the sensors consciously. For example, in Virtual Aquarium, some users imitated the movement of their toothbrushes in order to make the fishes dancing. There are two approaches to solve this problem. The first approach is to prohibit cheating by increasing the accuracy of the movement analysis. The second approach is to encourage the user not to cheat to use sensors. We are very interested in adopting the second approach in our future case studies. This approach requires the user to think about the merit behind the desirable lifestyle. How technologies can be used to encourage a user to think more deeply is a very challenging issue.

4.2   Persuasive Expression

The user study on Persuasive Art shows that users preferred the tree and the Mona Lisa over the abstract and the still life. The reason given was that more figurative paintings were considered to be more "intuitive". While any visual representation can be used to relay information, shapes that come with pre-attached meanings (e.g. "a tree withering is a negative thing") are more capable of evoking emotional engagement. It is therefore important to remember this third design principle when choosing a presentation metaphor. Tan and Cheok [18] showed that a real creature is found to arouse more empathy than a virtual creature. However, especially in Japan, people feel empathy also to virtual creatures. Fujinami [7] presented that Japanese users feel empathy for even virtual creatures represented as abstract symbols. We sometimes assign different meanings to a real creature and a virtual creature because we know the differences between them. We need to investigate the effect of virtual creatures as a persuasive expressions in future case studies.

In the future, it is necessary to consider how the feedback information appeals users without the explanation about the interpretation of the expressions because ambient lifestyle feedback information will appear anywhere to visualize a variety of aspects of our lifestyle. The metaphor visualizing a user's lifestyle helps him to notice the feedback information. The concept of affordance could be a guideline in designing linkages between activities and feedback. Product semantics [14] may be one suitable theory to help how feedback expression affords the meaning of the expression. A user sometimes mistakes to make the meaning of an expression, and this is one of the serious problem to rely on affordance. The user tends to define the non intentional meaning of an expression [2]. For example, an ugly picture may be used to discourage to keep the current undesirable habits, but the picture may encourage to keep the current undesirable habits for some avant-garde people. This is highly depending on the cultures and personalities of the users. It is not easy for a designer to assign a single meaning to a specific expression by all people. We believe that the expression of showing some virtual creatures is more acceptable to most of the people. Of course, each person may love different creatures. Empathetic feeling is a key to design successful ambient lifestyle feedback systems.

There are some very close systems to ambient lifestyle feedback systems. Playful toothbrush [3] shows a virtual teeth representing the current status of the user's toothbrushing. It explicitly shows the goals of the user's behavior. The user continues to use and enjoy the systems until he achieves the goal. However, motivating a user based on a long-term goal is important to maintain desirable lfestyle. The advanced motivation theories [16] help us to develop more effective case studies. It is useful to distinguish a short-term goal and a long-term goal to encourage the change away from undesirable habits and to keep desirable habits. In our case studies, Virtual Aquarium sets a short-term goal to complete sufficient toothbrushing in every night. On the other hand, EcoIsland sets a long-term goal to reduce CO_2 emission. Showing the explicit goal is effective to keep desirable habits until achieving the goal. The goal setting should be carefully designed so as not to stop desirable habits before achieving the goal. The combination of a short-term and a long-term goal is a very effective way to motivate a better lifestyle. Also, it is important to consider how to represent a goal in the expression. In Persuasive Art, the growth of the tree can be reinitialized every week to start a new goal, but it may reduce the sense of achievement in long-term. The relationship between the expression and goal setting should be investigated more in the near future.

4.3   Feedback Control and Emotion

In operant conditioning, feedback content can be divided into reinforcement and punishment depending on whether some behavior is encouraged or discouraged. Reinforcement and punishment are further divided into four types [4]:

- Positive reinforcement: encouraging the user's behavior by providing a favorable stimulus in response to it
- Negative reinforcement: encouraging the user's behavior by removing an averse stimulus in response to it
- Positive punishment: discouraging the user's behavior by providing an averse stimulus in response to it

– Negative punishment: discouraging the user's behavior by removing a favorable stimulus in response to it

Our case studies use the combination of the above four types of feedback. One of our finding is that the balance between positive reinforcement and positive punishment is very important in changing a user's behavior permanently. The user may be bored if the expression offers only positive reinforcement. On the other hand, the user may give up his hope to change undesirable habits when only positive punishment is offered. An appropriate balance is important in order to change the user's behavior permanently.

Jordan classified pleasure into four types: physio-pleasure, psycho-pleasure, socio-pleasure, ideo-pleasure [13]. This classification is a useful tool to design a reinforcement and a punishment. Physical comfort and discomfort are used as reinforcement and punishment to change the user's behavior. For example, a chair may change our comfortability by moving the backrest or arms. In the near future, we are interested in using physio-pleasure to design smart objects that change their shape according to the user's current behavior. In most of our case studies, the user's behavior is changed due to positive and negative emotion caused by the expression representing the user's current behavior. Dancing fishes make the user exited and increase his positive emotion, but when Mona Lisa is getting old, the negative emotion is increased and he feels anxious. Emotion is a very powerful tool to change a user's behavior [6] and we will try to develop a systematic way to use positive and negative emotions. We are going to enhance the use of social aspects into the feedback expression. If all people know the rules, the expression displayed in a public space would put interesting pressure on the user who is the target of the information. In EcoIsland, we have tried to use socio-pleasure as the feedback of the user's activities to reduce CO_2 emission. Social effects are an interesting tool in designing the feedback and need to be investigated more in the near future. Ideo-pleasure is interesting to be used to change the user's attitude in future case studies. The user's long-term good attitude will permanently change his undesirable habits and maintain his desirable habits. Ideo-Pleasure makes it possible motivate the user by himself. He has a belief called self-efficacy that he will be able to achieve his goal. Contemporary arts make us to think deeply about our future like sustainability and peace in the world. We like to consider how to use the expression of contemporary arts to persuade the user to change his attitude in future case studies.

4.4 Evaluating the Effects

We conducted a 8-12-day pilot study to show the effectiveness of Virtual Aquarium. The participants were three male and four female volunteers from three households associated with the researchers. All of them are Japanese. We recorded toothbrushing time per session for each participant changed during the study. In all cases, the brushing times were less than three minutes per session at the beginning of the study, and increased to over three minutes when the Virtual Aquarium was introduced. When the feedback system was removed the times fell back, yet mostly remained higher than their initial levels for the remainder of the study. How well the sensor data describes actual behavior obviously critically affects the validity of the results. Users could have cheated the system by waving the brush in their hands. Whether this should be seen as a problem depends on who is using the system on whom. In a concluding interview, the participants of this study self-reported that their brushing patterns had in fact changed, at least for the time being.

It is also worth mentioning that five of the users said that their families became interested in the status of the aquariums, e.g. following the changes in the number of fish. Virtual Aquarium, and by proxy the behavior of the user, became something of a topic of daily discussion in these families. We like to investigate to use ambient lifestyle feedback systems as social media that a user feels closer among her family members.

We believe that ambient lifestyle feedback systems will be accepted by most of people , and will become a part of our daily life to shape our lifestyle . We have presented more detailed evaluation results in [15].

5   Future Directions

In our daily lives, most of our behavior do not return adequate feedbacks. If computer technologies help to return adequate feedbacks to the user, they make him aware of his current lifestyle and he can change undesirable habits and maintain desirable habits easily. In the paper, we introduced a brief overview of ambient lifestyle feedback systems and four case studies. Each case study gave us various insights and we showed several findings with the case studies.

Economic benefit is a strong incentive to motivate the user to change his behavior. We have introduced the Eco credit concept in EcoIsland. The user will be encouraged if he has an incentive to get a return for his effort or contribution to reduce CO_2 emission. The user's activities are monitored by the system and paid to users in order to stimulate their desirable actions. In EcoIsland, a user can use the credit to purchase decorating items or to trade eco-unconscious activities. Thus, the user will be both a consumer and a producer. We believe that we can accelerate the money circulation by adding economical concepts in ambient lifestyle feedback systems.

There are many places to encourage a user to change his/her behavior to motivate a better lifestyle. In our case studies, Persuasive Art and Virtual Aquarium assume that the systems are installed in the user's house. In the near future, our daily life was become more nomadic and we often stay in hotels for personal or business reasons. In this case, we cannot see the status of the tree or the dancing the fishes. Some participants in the user study in Virtual Aquarium told that they wanted to take care of the fishes even when they were not stay at home. Thus, we believe that the feedback should be reflected in many places such as hotels and public spaces. One of the problem to realize the goal is that the user needs to find which digital expression reflects the feedback of his behavior. However, if we can use a public display to show the feedback of the user's behavior, it makes it possible to use social factors as positive reinforcement and punishment. Ambient lifestyle feedback systems may be installed everywhere to enhance a variety of our daily activities in future, but we also need to consider whether this is our dream or just a nightmare. Is this really a better lifestyle for the future ? Also, using ambient lifestyle feedback systems everywhere may take control of our attitude, which may cause serious ethical problems. The user should have a right to control which behaviors are reflected in expressions. We also need to discuss who chooses expressions to reflect the user's behavior. Does the system choose expressions automatically ? Can the user choose expressions manually ? It is important to consider how to reflect the user's behavior when trying to change multiple behaviors at the same time.

References

1. Asch, S.E.: Opinions and social pressure. Scientific American, 31–35 (1955)
2. Brandes, U., Erlhoff, M.: Non Intentional Design, daad (2006)
3. Chang, Y., Lo, J., Huang, C., Hsu, N., Chu, H., Wang, H., Chi, P., Hsieh, Y.: Playful toothbrush: UbiComp technology for teaching tooth brushing to kindergarten children. In: Proceedings of ACM CHI 2008 (2008)
4. Domjan, M.P.: The Principles of learning and behavior, 5th edn., Wadsworth (2002)
5. Fogg, B.J.: Persuasive technology: using to change what we think and do. Morgan Kaufmann, San Francisco (2002)
6. Fredrikson, B.L.: The Value of Positive Emotions: The emerging science of positive psychology is coming to understand why it's good to feel good. American Scientist 91 (July-August 2003)
7. Fujinami, K., Riekki, J.: A Case Study on an Ambient Display as a Persuasive Medium for Exercise Awareness. In: Proceedings of the 3rd International Conference on Persuasive Technology (2008)
8. Hallnas, L., Redstrom, J.: Slow technology - designing for reflection. Personal and Ubiqutious Computing 5(3) (2001)
9. Hardin, G.: The tragedy of the commons. Science 162, 1243–1248 (1968)
10. IPCC. IPCC Fourth Assessment Report: Climate Change (2007), http://www.ipcc.ch/
11. Nawyn, J., Intille, S.S., Larson, K.: Embedding behavior modification strategies into a consumer electronics device. In: Dourish, P., Friday, A. (eds.) UbiComp 2006. LNCS, vol. 4206, pp. 297–314. Springer, Heidelberg (2006)
12. Jafarinaimi, N., Forlizzi, J., Hurst, A., Zimmerman, J.: Breakway: An ambient display designed to change human behavior. In: Proceedings of ACM CHI 2005 (2005)
13. Jordan, P.W.: Designing pleasurable products: An introduction to the new human factors, Routledge (2002)
14. Krippendorff, K.: The semantic turn: an new foundation for design. CRC Press, Boca Raton (2005)
15. Nakajima, T., Lehdonvirta, T.V., Tokunaga, E., Kimura, H.: Reflecting Human Behavior to Motivate Desirable Lifestyle. In: Proceedings of ACM Designing Interactive Systems 2008 (2008)
16. Reeve, J.: Understanding Motivation and Emotion, 4th edn. Wiley, Chichester (2005)
17. Rohs, M.: Visual code widgets for marker-based interaction. In: Proceedings of ICDCS Workshops: IWSAWC 2005 (2005)

18. Tan, R.K.C., Cheok, A.D.: Empathetic Living Media. In: Proceedings of ACM Designing Interactive Systems 2008 (2008)
19. Weiser, M., Brown, J.S.: Designing calm technology (December 21, 1995) [online, February 27, 2007] ,
    http://www.ubiq.com/hypertext/weiser/calmtech/calmtech.htm
20. Zajonc, R.B.: Social facilitation. Science 149, 269–274 (1965)
21. List of diet and nutrition software for Palm [online, February 27, 2007],
    http://www.handango.com/
    SoftwareCatalog.jsp?siteId=1&platformId=1&N=96804+95712
22. List of quit smoking meters [online, February 27, 2007],
    http://www.ciggyfree.com/ODAT/
    index.php?option=com_content&task=view&id=205&Itemid=25

Raising Awareness about Space via Vibro-Tactile Notifications

Andreas Riener and Alois Ferscha

Johannes Kepler University Linz, Institute for Pervasive Computing,
Altenberger Str. 69, A-4040 Linz, Austria
Tel.: +43/7236/3343–920, Fax. +43/732/2468–8426
{riener,ferscha}@pervasive.jku.at

Abstract. Human perception, in a world of continuous and seamless exposure to visual and auditory stimuli, is increasingly challenged to information overload. Among the primary human senses, vision, audition and tactation, particularly the sense of touch appears underemployed in todays designs of interfaces that deliver information to the user. While about more than 70% of the information perceived by humans is delivered via the sight and hearing channel, only about 21% is perceived via the haptic sense. In situations of work or engaged activity, where both the visual and auditory channel are occupied because of the involvement in the foreground task, notifications or alerts coming from the background, and delivered via these channels tend to fail to raise sufficient levels of attention.

With this paper we propose to involve the haptic channel for situations where important notifications tend to be "overseen" or "overheard". We opt for a vibro-tactile notification system whenever eyes, ears and hands are in charge. A body worn, belt like vibration system is proposed, delivering tactile notifications to the user in a very subtle, unobtrusive, yet demanding style. Vibration elements seamlessly integrated into the fabric of an off-the-shelf waist belt, lets the system deliver patterns of vibration signals generated by modulating amplitude, frequency, duration and rhythm – so called tactograms – to eight well positioned vibra elements. A series of user tests has been conducted, investigating the perception of distance to physical objects, like walls or obstacles, in the vicinity of users. Results encourage for a whole new class of space awareness solutions.

Keywords: Human Perception, Vibro-tactile Interfaces, Information Overload, Space Awareness.

1 Raising Human Attention

Traditional interface design involving human perception has mainly focussed on two modalities: *vision* and *hearing*. In many situations, however, when the vision and audition channel of perception are highly in charge, alerting and notifying important information can suffer from inattentiveness due to information overload. In such situations, to raise human attention, the additional use of *haptic*

D. Roggen et al. (Eds.): EuroSSC 2008, LNCS 5279, pp. 235–245, 2008.

sensation offers great potential for providing a supplementary level of awareness (about 21% of information is perceived with the haptic sense [1]; according to Mauter and Katzki [2] touch is used to a lower proportion, as 80% to 90% of all sensory input is received with the eyes). Steadily increasing information emergence directs to users oversaturation and distraction and encourages for new input and output modalities in human-centered computing. Other aspects which can generate added values in specific applications are the compensation of restrictions/limited awareness of visual and auditive sensations in noisy, dark or foggy environments.

Involving haptics in the design of user interfaces, particularly for the purpose of raising human attention in certain situations, or even to keep the user subtly informed about the environment has been addressed in the literature. Brewster and Brown [3] gives suggestions for using tactile output as display to enhance desktop interfaces and situations with limited or unavailable vision. Lindeman *et al.* [4], [5] states basic principles for haptic feedback systems and describes implementation-advances for their vibrotactile system "TactoBox". Additionally, the integration into a VR system (immersive simulator) of the U.S. Naval Research Laboratory has been depicted. Tsukada and Yasumara introduced in [6] a wearable tactile display called "ActiveBelt" for transmitting directional cues – one application ("FeelNavi") tries to map distance information to one of four pulse intervals. "feelSpace"[1] is a research project with the aim to investigate the effects of long-time stimulation with orientation information. On their belt, the element pointing north is always vibrating slightly, so that the person wearing the belt gets permanent input about his/her orientation. Although these systems presents novel approaches, none of them give qualitative evidence regarding user perception of haptic distance awareness.

Erp *et al.* uses haptic directional information for navigational tasks [7]. They mentioned that direction information alone is sufficient for the considered application, but distance information may be beneficial. In their walking experiments with haptic distance notifications if a certain test person comes within a range of 20 meters to the end point they coded distance into the proportion between signal to pause times, while vibration intensity is fixed. Tscheligi *et al.* are studying navigation support for pedestrians [8], which are often occupied with one or more other tasks. Therefore, visual data presentation on a electronic device like a PDA is not appropriate, nor is it auditive feedback because of environmental noise or conversations. As result, they see great potential for future navigation devices in unobtrusive systems based on tactile feedback (like the one presented in [7]), even if they are still under development. "Shoogle" [9] is a more recent approach for presenting eyes-free, single-handed interaction for mobile devices (data input with inertial sensors, realistic user feedback is given by stressing the two modalities vision and sound).

As possible platform for a vibro-tactile belt display, the QBIC-platform, presented in [10], could serve. Prewett *et al.* [11] presents a vibro-tactile display

---

[1] The study project feelSpace, URL:`http://feelspace.cogsci.uos.de` (April 12, 2008).

system to support the perception of the presence of the target interaction object in a 3D VR environment.

This experiment is the first in a larger series of studies considering haptic interaction, with the overall aim to investigate on the correlation between various haptic stimulations and corresponding user perception. We are interested in (*i*) distance- and position awareness, (*ii*) orientation awareness, (*iii*) awareness of danger objects or zones, (*iv*) mapping of important activities and (*v*) awareness about size and shape of objects. Here, only the aspect of *distance awareness* was considered, evaluations regarding quality, accuracy, reliability and users' personal opinions were given.

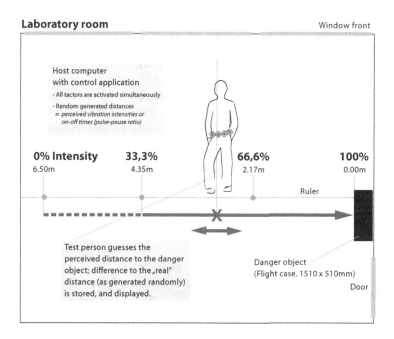

Fig. 1. Experimental setting for studying vibro-tactile distance-awareness

2 Space Awareness

For transmitting haptic stimulations to persons, a vibro-tactile belt system has been developed. It has the capability of generating intuitive room or space awareness by individually activating 8 tactor-elements around the waist (uniformly distributed), and, by variation of vibration intensity or frequency, allows for indicating object distances. The mapping between an object in space, and the corresponding activated tactors on the waist belt, is performed by a combination of software and hardware components, which are described briefly in the next section.

2.1   A Vibro-Tactile Waist Belt

The vibro-tactile waist system consists of the belt itself (with embedded tactor elements), the communication/control hardware, and a battery pack for mobile operation. On the host side, the framework is responsible for mapping information like distance, orientation, etc. to tactor control commands and for triggering the *haptic events*. Communication between the 130 x 90 x 45mm-sized belt hardware (board-size without housing is 74 x 93 x 33mm) and the host is established via Bluetooth connection. The weight of the entire system (including power supply and belt) is 680g, which is rather heavy but could be downsized by factor 2–3 when using a miniaturized, optimized setting (reference design for technical feasibility is given e.g. in [10] – the mainboard for the QBIC-platform is 44 by 55mm in dimension). Essential components of the system are an Atmel AVR Mega 32 Microcontroller for controlling the tactor elements, a ULN2803AG high-current darlington transistor array (8 pairs of NPN darlington transistors) for driving the vibration motors, a OEMSPA441 connectBlue industrial Bluetooth 2.0 serial port adapter (SPA), and 8 pieces of standard, low-cost Nokia 7210 cell-phone vibramotors, housed in a PVD-contactor with a contact area of approximately 3 by 2cm. The wireless system is completed by a PM85-22 universal rechargeable 22Wh Li-ion battery pack. The software part for generating random values, translating them into appropriate control commands and delivering them wirelessly to the vibro-tactile belt has been implemented as a combination of JAVA/C++ applications, exchanging information with the Atmel microcontroller via Bluetooth communication.

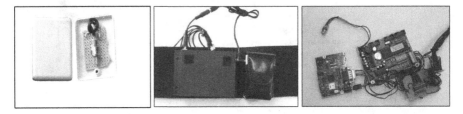

Fig. 2. Vibration elements, belt hardware with battery pack, and control hardware, dismantled from the housing (from left to right)

Tactor Placement: A good starting point for considerations about the where to place tactors can be found in [4, p.147f]. For the experiment, only the region of waist is significant. Weinstein [12] investigated the spatial resolution of receptors in humans' skin. He found that the threshold distance in the region of the abdomen is approximately $30mm$. The calculation of the minimum number of tactor elements around the waist can be derived from the waist-circumference of average males and females, which is between 86 and 89cm for men, and between 74 and 81cm for woman (with an overall mean of 82.5cm).

$$\frac{825mm \text{ [mean waist-circumference]}}{2 * 30mm \text{ [threshold diameter]}} = 13.75 \text{ [tactors]}$$

According to this estimation, the maximum accuracy for (orientation) awareness on a vibro-tactile belt system should be achieved when placing 14 tactor elements around the waist. As we only consider distances in the present experiment, the utilized 8-tactor belt system should be sufficient to provide complete sensations around the waist.

2.2   Tactor Actuation

Our vibro-tactile belt system is responsible for haptic feedback based on 4 parameters: (i) vibration intensity (pulse-width modulation, PWM), (ii) activation frequency, (iii) duration (relationship between pulse- and pause-times), and (iv) rhythm. The dynamic behaviour of all vibro-tactile elements (*tactors*) in a system is specified by vibro-tactile patterns or "*tactograms*". For the current studies, all parameters except the fourth (*rhythm*) had been utilized and diversified.

Vibration intensity: The intensity of vibro-tactile transducers is controlled with pulse-width modulation (PWM) in time-per-unit. The implementation of the hardware controllers used in the present experiment allows only one shared PWM-value for all tactor elements (which is sufficient because of equal actuation of all tactors). Individual intensities per vibration element would be possible by variation of the stimulation frequency, which would be perceived in a similar manner than changes in the amplitude.

Activation frequency: In our setting, vibration frequency is individually adjustable for a certain tactor. For the current experiments all tactors are switched on and off simultaneously. As we didn't evaluate orientation information, the activation all around the waist was used to get a higher perception of vibro-tactile distance information. Vibration intensity and frequency were choosen according to the functional features of pacinian corpuscles or mechanoreceptors (they are embedded in almost the entire body and adapt rapidly to intensity changes; furthermore they are sensitive in the frequency range from 10 to $500hz$, with highest perception at $\approx 250hz$ [13], [14], [15]).

Pulse-pause time (duration): The variation parameter *interval length* was used as another alternative to notify about changing distances between a person and an obstacle. The maximum interval length was selected according to several investigations on response times for HCI-systems, which recommends a maximum response time of two seconds or less (see Testa *et al.* [16, p.63], Miller [17], or Shneiderman *et al.* [18, Chap.11]).

3   Experimental Design

The vibro-tactile waist belt, which is adjustable in length to enable a tight fit, was attached and adjusted to each of the test participants (ranging from 23 to 32 years in age). All of them had been informed before the test, to wear only a thin shirt for better vibration perception. Prior the recording of real measurements we did a *calibration task* by transmitting a number of given distances (variation of the parameters *vibration intensity* and *interval length*) to the test candidates (0% or 650.00*cm*, 33% or 435.00*cm*, 66% or 217.00*cm*, and 100% or 0.00*cm*, see Fig. 1). After that, for each test person a series of consecutive random intensity-values had been generated and wirelessly delivered to the belt – the users estimate for the distance, together with the system-generated distance value, were stored in a database.

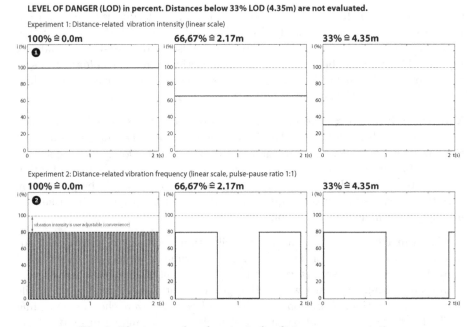

Fig. 3. The two analzyed variants for distance representation

In first user tests with the proposed system we experienced, that low-intensity vibrations (below 33%) were almost indistinguishable. For that reason, the final system configuration had been changed to use only vibration intensities between 33% and 100% (or distance values between person and obstacle of below 4.35*m*, as depicted in Fig. 3). For comparability issues, this definition has also been applied to the second experiment on variation of vibration frequency.

3.1 Experiment 1: Distance Representation by Variation of Vibration Intensity

In this experiment series, the distance between an obstacle and a test participant was mapped to the vibration intensity of a vibro-tactile actuator. The vibration was initiated on all tactor elements simultanously (this generates a higher perception than activation only a single vibrating element). For the mapping between distance and vibration intensity, a linear correlation function between distance and vibration intensity has been used (inversely proportional). The *level of danger* (LOD) for a person, and thus, the vibration intensity, increases with declining distance from the danger object (see top row in Fig. 3).

The entire setting has been installed in a laboratory room, as shown in Fig. 1. Test candidates had been briefed to move forewards and backwards, alongside a floor-mounted ruler, and transmit their opinion for the felt distance of a specific vibration-intensity value verbally. We conducted two different types of runs, the first with no feedback on the real distance values, and another with a direct feedback on estimated values' real counterpart.

3.2 Experiment 2: Distance Representation by Variation of Vibration Frequency

In a second test series we experimented with variation of the interval length for representing changing distances. This approach follows the idea of Park Distance Control (PDC) systems, established in the automotive domain. A PDC is a distance warning system that provides information on the distance to the nearest obstacle. The frequency of the signal increases as the distance to the obstacle decreases, a continuous signal is output in very close proximity to obstacles (30cm or less) [19, p.4].

The vibration strength was adjusted individually for each test person to be of convenient intensity, and was than fixed for the entire experiment. The ratio between pulse- and pause times was set to 1 for the whole experiment. A distance of *zero* between person and object (represented by a minimum value of 0.05m) was mapped to a (near-continuous) vibration with a frequency of 43.5hz (according to the formula below), the maximum distance (according to the first experiment at the LOD-level of 33%, or 4.35m) between danger object and person was represented by a vibration frequency of 0.5hz (this implies a maximum interval length of 2 seconds).

$$\text{pulse time [ms]} = \frac{d \, [\text{distance in } m] * 2,000ms \, [\text{max. interval length}]}{2 * 4.35m \, [\text{max. distance}]}$$

For this experiment, again the two variants (i) without feedback, and (ii) with immediate feedback have been tested. The experiment itself was designed and conducted about 3 month after the first experiment, a subset of the test participants from the first test attended this second experiment series (because of the long time between the two experiments we can almost neglect possible dependencies of the second experiment from the first one).

4   Evaluation

Experiment 1: Variation of vibration intensity: The mean value of estimation errors is 0.438m for the experiment without feedback, and 0.299m for the second case with immediate feedback (see Table 1, and charts (1) and (3) in Fig. 4). This is a qualitative improvement of the second setting of 46.49% and confirms the assumption that *training* or *learning* is essential for a reliable

Table 1. Distance representation by variation of vibration intensity: No feedback (left), immediate feedback (right). Data represents the deviations from the real distance in meters.

Test Person x_{min}	Min x_{min}	Max x_{max}	Mean $\overline{x}$	Median $\widetilde{x}$	Std.Dev. σ	Min x_{min}	Max x_{max}	Mean $\overline{x}$	Median $\widetilde{x}$	Std.Dev. σ
	Series without feedback					Series with immediate feedback				
1	0.005	1.070	0.346	0.315	0.267	0.030	0.710	0.313	0.315	0.193
2	0.040	0.960	0.390	0.270	0.290	0.005	1.130	0.357	0.255	0.296
3	0.055	1.470	0.563	0.370	0.412	0.015	0.735	0.250	0.255	0.207
4	0.010	1.295	0.470	0.455	0.358	0.000	0.830	0.307	0.220	0.258
5	0.030	1.610	0.734	0.660	0.470	0.030	0.950	0.317	0.255	0.219
6	0.010	0.805	0.261	0.195	0.214	0.035	0.665	0.295	0.270	0.191
7	0.015	0.990	0.487	0.440	0.267	0.010	0.805	0.301	0.250	0.253
8	0.005	1.240	0.455	0.355	0.368	0.045	0.435	0.247	0.260	0.137
9	0.000	0.835	0.351	0.255	0.282	0.025	0.755	0.334	0.255	0.217
10	0.000	0.985	0.382	0.350	0.272	0.025	0.820	0.326	0.290	0.206
All	**0.000**	**1.610**	**0.438**	**0.363**	**0.350**	**0.000**	**1.130**	**0.299**	**0.255**	**0.223**

user interface. This tendency is valid for all test candidates, except participant number 6 – here the statistical results are better for the feedback-free first experiment, although the maximum deviation error $(0.805m)$ is somewhat higher in the first experiment $(0.665m)$. This could be interpreted either as higher tactile sensibility of the person or simply as statistical outlier.

The left column in Fig. 4 shows the corresponding deviation errors in meters for both experiment variants. Dotted and dashed lines indicates the linear tendencies for the deviation error. They suggest the assumption that the error increases with number of experiments in the first case (without feedback), and decreases for the second case with permanent feedback (the slope of the trend lines is more pronounced when viewing the charts for individual test attendees). To find the upper and lower convergence boundaries for the error values, larger test series with 100+ readings for both experiments would have to be accomplished.

During experimentation we perceived several qualitative remarks from the test candidates:

(i) Not only the felt vibration, but also the "noise" caused by this vibration, has been used as information channel for distance-perception.
(ii) The self-assurance with estimating distances increased with increasing duration of the experiment.
(iii) The first series of experiments (without any feedback) provoked less cognitive load.
(iv) A major variation in vibration intensity, e.g. a high vibration value followed by a low one and vice versa, makes it harder to estimate the corresponding distance (consequently, such variations results in higher estimation errors).

Experiment 2: Variation of vibration frequency: The results of this experiment contrast the ones from the first experiment.

Table 2. Distance representation by variation of vibration frequency: No feedback (left), direct feedback (right). Data represents the deviations from the real distance in meters.

Test Person	Min x_{min}	Max x_{max}	Mean $\overline{x}$	Median $\widetilde{x}$	Std.Dev. σ	Min x_{min}	Max x_{max}	Mean $\overline{x}$	Median $\widetilde{x}$	Std.Dev. σ
	Series without feedback					Series with immediate feedback				
1	0.000	1.670	0.529	0.385	0.491	0.010	1.040	0.401	0.380	0.264
2	0.040	1.220	0.428	0.295	0.314	0.010	1.650	0.489	0.375	0.425
3	0.050	1.860	0.588	0.480	0.458	0.010	1.030	0.355	0.330	0.277
4	0.010	1.130	0.391	0.245	0.329	0.000	1.230	0.325	0.225	0.292
5	0.030	2.600	0.528	0.445	0.518	0.010	1.140	0.433	0.395	0.317
All	**0.000**	**2.600**	**0.492**	**0.380**	**0.430**	**0.000**	**1.650**	**0.401**	**0.370**	**0.321**

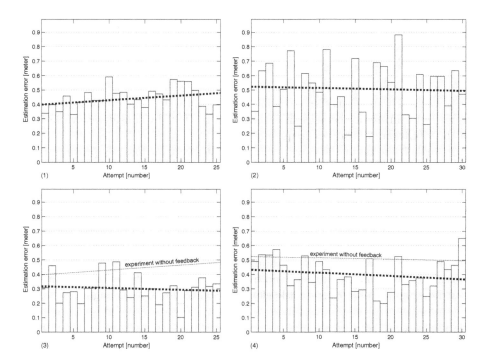

Fig. 4. Plot of averaged estimated values and their corresponding errors for both styles of the conducted experiments, "level of danger according to vibration intensity" (left column) respectively "vibration frequency" (right colum). The top row represents the test series with no feedback, the bottom row depicts the test series with immediate feedback. Dotted and dashed lines shows the deviation-tendency (linear trend lines).

Mean estimation errors are $0.492m$ (no feedback), respectively $0.401m$ (immediate feedback), see Table 2. The improvement is with 22.69% below the half of the improvement of experiment one (which was 46.49%). Comparing the absolute estimation errors we can clearly see that the second experiment performs worse compared to the first: A mean estimation error of $0.492m$ compared to

$0.438m$ stands for a degradation of 12.3% for the feedback-free case, $0.401m$ compared to $0.299m$ means a degradation of 34.11% for the second case with direct feedback.

Furthermore, the chart of estimations with associated errors (right column in Fig. 4) shows, that the trendlines are nearly flat (gently sloping). This encourages the assumption that a decrease in estimation error is hardly possible by training for this kind of stimulation.

Most of the experiment attendees (which were the same as in the first experiment) reported, that distance estimation from the experimental system with varying the vibration intensity was more intuitive than the second setting with varying the vibration frequency. Evaluation results confirmed these statements (see Table 2, and charts (2) and (4) in Fig. 4).

5   Conclusions

Motivated by the aim for raising human attention, or subtly delivering information about a users environment aside the vision and hearing channel of perception, we have developed a vibro-tactile distance awareness system, implemented as a waist belt. User studies confirmed our hypothesis, that physical distance, to walls, furniture or obstacles in general, can be "felt" by considerate variations of the vibration intensity or frequency.

A further finding of the experiments is, that direct feedback improves on the estimation precision.

The results encourage for a whole new class of pervasive and ubiquitous computing applications and user interface designs, that lets users "feel" the environment they are in.

Above these results, we expect further improvements on haptic distance awareness by (i) integrating notions of orientation and direction, (ii) further experiments with different models of distance and vibration-intensity mapping, e.g. non-linear variations of intensity and frequency, and (iii) additional experiments based on high-quality vibro-tactile elements to avoid faults and signal noise.

References

1. Schmidt, R.F., Lang, F., Thews, G.: Physiologie des Menschen mit Pathophysiologie (Springer Lehrbuch) (Gebundene Ausgabe). Springer, Berlin (2005)
2. Mauter, G., Katzki, S.: The Application of Operational Haptics in Automotive Engineering. In: Business Briefing: Global Automotive Manufacturing & Technology 2003, Team for Operational Haptics, Audi AG, pp. 78–80 (2003)
3. Brewster, S., Brown, L.M.: Tactons: structured tactile messages for non-visual information display. In: AUIC 2004: Proceedings of the fifth conference on Australasian user interface, Darlinghurst, pp. 15–23. Australian Computer Society, Inc., Australia (2004)
4. Lindeman, R.W., Page, R., Yanagida, Y., Sibert, J.L.: Towards full-body haptic feedback: the design and deployment of a spatialized vibrotactile feedback system. In: VRST 2004: Proceedings of the ACM symposium on Virtual reality software and technology, pp. 146–149. ACM, New York (2004)

5. Lindeman, W., Yanagida, Y., Noma, H., Hosaka, K.: Wearable vibrotactile systems for virtual contact and information display. Virtual Real 9(2), 203–213 (2006)
6. Tsukada, K., Yasumura, M.: ActiveBelt: Belt-Type Wearable Tactile Display for Directional Navigation. In: Davies, N., Mynatt, E.D., Siio, I. (eds.) UbiComp 2004. LNCS, vol. 3205, pp. 384–399. Springer, Heidelberg (2004)
7. Van Erp, J.B.F., Van Veen, H.A.H.C., Jansen, C., Dobbins, T.: Waypoint navigation with a vibrotactile waist belt. ACM Trans. Appl. Percept. 2(2), 106–117 (2005)
8. Tscheligi, M., Sefelin, R.: Mobile navigation support for pedestrians: can it work and does it pay off? Interactions 13(4), 31–33 (2006)
9. Williamson, J., Murray-Smith, R., Hughes, S.: Shoogle: excitatory multimodal interaction on mobile devices. In: Proceedings of the SIGCHI conference on Human factors in computing systems (CHI 2007), pp. 121–124. ACM, New York (2007)
10. Amft, O., Lauffer, M., Ossevoort, S., Macaluso, F., Lukowicz, P., Troster, G.: Design of the QBIC Wearable Computing Platform. In: ASAP 2004: Proceedings of the Application-Specific Systems, Architectures and Processors, 15th IEEE International Conference, Washington, DC, USA, pp. 398–410. IEEE Computer Society, Los Alamitos (2004)
11. Prewett, M.S., Yang, L., Stilson, F.R.B., Gray, A.A., Coovert, M.D., Burke, J., Redden, E., Elliot, L.R.: The benefits of multimodal information: a meta-analysis comparing visual and visual-tactile feedback. In: Proceedings of the 8th international conference on Multimodal interfaces (ICMI 2006), pp. 333–338. ACM, New York (2006)
12. Weinstein, S.: Intensive and extensive aspects of tactile sensitivity as a function of body part, sex, and laterality. In: Kenshalo, D. (ed.) The skin senses, Springfield, pp. 195–218 (1968)
13. Bolanowski, S., Gescheider, G., Verrillo, R., Checkosky, C.: Four channels mediate the mechanical aspects of touch. Acoustical Society of America Journal 84, 1680–1694 (1988)
14. Gibson, R.H.: Electrical stimulation of pain and touch. In: Kenshalo, D.R. (ed.) First International Symposium on Skin Senses, Florida State University, Tallahassee, Springfield, Illinois, Charles C. Thomas, March 1968, pp. 223–261 (1966); WR 102 I61
15. Pohja, S.: Survey of Studies on Tactile Senses. ISSN : 0283-3638 R960:02, RWCP Neuro Swedish Institute of Computer Science (SICS) Laboratory, Box 1263, S-164 28 Kista, Sweden (March 1996) ISSN: 0283-3638
16. Testa, C., Dearie, D.: Human factors design criteria in man-computer interaction. In: ACM 1974: Proceedings of the 1974 annual conference, pp. 61–65. ACM, New York (1974)
17. Miller, R.B.: Response time in man-computer conversational transactions. In: Proceedings of AFIPS Fall Joint Computer Conference, vol. 33, pp. 267–277 (1968)
18. Shneiderman, B., Plaisant, C.: Designing the User Interface: Strategies for Effective Human-Computer Interaction., 4th edn. Pearson Education, Inc., Addison-Wesley Computing (2005), ISBN: 0-321-19786-0
19. BMW: 2007 BMW E70 X5 Reference Documents – Park Distance Control (PDC). Technical report, p. 19 (October 2006),
http://www.x5world.com/pdf/e70/05c1_E70ParkDistanceControl.pdf

Author Index